Food Security Governance in the Arctic-Barents Region

Kamrul Hossain • Dele Raheem • Shaun Cormier

Food Security Governance in the Arctic-Barents Region

 Springer

Kamrul Hossain
NIEM
University of Lapland NIEM
Rovaniemi, Finland

Dele Raheem
NIEM
University of Lapland NIEM
Rovaniemi, Finland

Shaun Cormier
NIEM
University of Lapland NIEM
Rovaniemi, Finland

ISBN 978-3-030-09319-8   ISBN 978-3-319-75756-8   (eBook)
https://doi.org/10.1007/978-3-319-75756-8

Printed on acid-free paper

This Springer imprint is published by the registered company Springer International Publishing AG
part of Springer Nature.
The registered company address is: Gewerbestrasse 11, 6330 Cham, Switzerland

Preface

This book aims to fill gaps in the knowledge about food security and the coordination of a legal framework for its governance in the context of the Arctic-Barents region. The Food and Agriculture Organization of the United Nations (FAO 1996) stated that "food security exists when all people at all times have physical, social and economic access to sufficient, safe and nutritious food to meet their dietary needs and food preferences for an active and healthy life." However, the authors of this book perceive a wider context, in which "food sovereignty" is an integrated part of food security. Food sovereignty highlights the food preferences of the consumers. In the Arctic-Barents region, traditionally available foods are generally preferred among the communities. The consumption of these foods includes traditional and local foods, for example, from reindeer herding, hunting, fishing and berry picking. These foods can be readily accessed, and they are shared among the communities in the region. These foods also arguably meet the required dietary needs for the healthy and active lives of the individuals in the communities. In addition, the practice around the consumption of traditional foods is also tied with many rituals of traditional communities, such as those of indigenous peoples. Therefore, having a say in regard to food preference promotes additional value to local food, which eventually ensures "… the ability and the right of people to define their own policies and strategies for the sustainable production, distribution and consumption of food that guarantee the right to food for the entire population, …" as referred to by the World Forum on Food Sovereignty (WFS 2001).

The proposed book will form part of the main outcome from the Academy of Finland's ongoing project on Human Security as a Promotional Tool for Societal Security in the Arctic: Addressing Multiple Vulnerability to its Population with Specific Reference to the Barents Region (HuSArctic). By promoting food security will enrich the closely overarching concept of human security, which is to protect the vital core of all human lives in ways that enhance human freedom and fulfilment. Food security and safety will be better improved in the Arctic-Barents region by implementing relevant policies at all levels of government. A variety of challenges remain in many Arctic-Barents communities regarding food security; it is necessary to promote research and prioritize policies that will promote food sovereignty, as it arguably coincides with food security in the region. The empowerment from food sovereignty will further help people to identify with their culture and natural environment and enhance their knowledge about traditional food systems that can improve health and build community support. In the Arctic-Barents region, food is a way of life for many people; it helps people realize the importance of maintaining their connections with nature and their own cultures, and between heart and mind, to reaffirm identity. Food is important for both indigenous and non-indigenous peoples; the use and consumption of traditional and local foods will help unite and connect people and their identities, traditions and cultures within and between communities.

This book, while providing general reference to the Arctic region, will focus specifically on the Arctic-Barents region, an area of northern Europe containing Norway, Sweden, Finland and Russia, with approximately 5.2 million indigenous and non-indigenous peoples. The Saami, Veps, Komi, Nenets and Pomors, along with the non-indigenous peoples of the region, transcend across these national and international boundaries to maintain their traditional activities associated with food. Some of these activities include herding, fishing, foraging and hunting, which are all practiced in cooperation with family members in group atmospheres and in their communities. Food remains a vital component in the lives of the people living in this region. However, it is hampered by regional transformations resulting from climate change and gradually increased human activities, such as mining, oil and gas developments, tourism and shipping. Industrial food processing has been identified as a major contributor to global warming and market foods often have to cover long distances to reach consumers, and as such they need to be well preserved and protected by packaging materials. Market food prices are often dictated by the supply and demand from the international market. The limited availability and accessibility of local foods makes people rely on store-bought foods that have been transported over long distances. Food miles are linked with growing concerns about the emissions of carbon dioxide and other greenhouse gases from fossil fuel-based transport. Hence, the promotion of local and traditional foods helps ensure food security and helps reduce carbon in the region's food system.

To date, most of the available studies on traditional food systems and their impacts on food security are from the Canadian, Greenland and US Arctic regions. These circumpolar regions are similar to the Barents European Arctic in terms of geographically having a heavy presence of indigenous inhabitants. By focusing on the Arctic-Barents region, this book offers a more balanced and systemic review on the role of traditional foods in the less explored Barents-Arctic communities. We have used the term Arctic-Barents region synonymously with the Euro-Arctic-Barents region discussed in this book—referring to the region of the European High North, which is characterized by high latitude, a circumpolar location and pristine nature.

This book, while highlighting the detailed and all-encompassing picture related to food security in the regional context, offers an analysis of the existing regulatory and policy tools in connection with the governance of food security in regional settings. Hence, it presents a number of recommendations based on identified gaps in order to promote overall food security in the region. It is envisaged that relevant actors as well as other stakeholders will find the book to be an important contribution to the promotion of policies and strategies on food security.

Rovaniemi, Finland   Kamrul Hossain
                                                                                                          Dele Raheem
                                                                                                      Shaun Cormier

Acknowledgements

We are grateful to all those with whom we have had the pleasure of working with while accomplishing this task. During this process, a number of academic scholars as well as other stakeholders have extended their generous support in terms of providing us with new knowledge. We particularly thank the following scholars at the Arctic Centre of the University of Lapland, who have given their time freely for interviews on topics related to their fields of work—Professors Bruce Forbes and Monica Tennberg and Senior Researcher Juha Joona. We also gratefully acknowledge Dr. Päivi Soppela's contribution in providing us with some of the pictures used in this book. Finally, we thank our colleague Ms. Anna Petrétei, who has been so kind to read over the manuscript diligently as a final check for any technical errors.

This book would not have been possible without the generous financial support granted by the Academy of Finland within the framework of the research project Human Security as a Promotional Tool for Societal Security in the Arctic: Addressing Multiple Vulnerability to its Population with Specific Reference to the Barents Region (HuSArctic), hosted at the Northern Institute for Environmental and Minority Law at the Arctic Centre of the University of Lapland. We are therefore thankful to both the Academy of Finland and the Arctic Centre of the University of Lapland.

Abbreviations

AC	Arctic Council
ACAP	Arctic Contaminants Action Program
AMAP	Arctic Monitoring and Assessment Programme
BEAC	Barents Euro-Arctic Council
BEAR	Barents Euro-Arctic Region
BWC	Ballast Water Convention
CAC	Codex Alimentarius Commission
CAFF	Conservation of Arctic-Barents Flora and Fauna
CBD	Convention on Biological Diversity
COP	Committee of Parties
CVD	Cardio-vascular diseases
DDE	Dichlorodiphenyldichloroethylene
DDT	Dichlorodiphenyltrichloroethane
EFSA	European Food Safety Authority
EPPR	Emergency Prevention, Preparedness and Response
EU	European Union
EVIRA	Elintarviketurvallisuusvirasto (Finnish Food Safety Authority)
FAO	Food and Agriculture Organization
FFN	Food Fraud Network
GHG	Greenhouse gases
HCB	Hexachlorobenzene
ICCPR	International Covenant on Civil and Political Rights
ICESCR	International Covenant on Economic, Social and Cultural Rights
ILO	International Labour Organization
IMO	International Maritime Organization
LRTAP	Long-Range Trans-Boundary Air Pollution
MARPOL	International Convention for the Prevention of Pollution from Ships
MDG	Millennium Development Goals
ND	Northern Dimension
NFA	National Food Administration, Sweden (Livsmedelsverket)
OHCHR	Office of the United Nations High Commissioner for Human Rights

OSPAR	Convention for the Protection of the Marine Environment of the North-East Atlantic
PAME	Protection of the Arctic-Barents Marine Environment
PCB	Polychlorinated biphenyls
PFAS	Perfluorinated alkylated substances
PFDA	Perfluorodecanoic acid
PFHpS	Perfluoroheptane sulfonate
PFHxS	Perfluorohexane sulfonic acid
PFNA	Perfluorononanoic acid
PFOA	Perfluorooctanoic acid
PFUnDA	Perfluoroundecanoic acid
POP	Persistent organic pollutants
SDWG	Sustainable Development Working Group
SPS	Sanitary and Phytosanitary
SOLAS	Safety of Life at Sea
UNCLOS	United Nations Convention on the Law of the Seas
UDHR	Universal Declaration of Human Rights
UNDRIP	United Nations Declaration on the Rights of Indigenous Peoples
UNECE	United Nations Economic Commission for Europe
UNEP	United Nations Environment Programme
UNFCCC	United Nations Framework Convention on Climate Change
WIPO	World Intellectual Property Organization
WTO	World Trade Organization
WWF	World Wide Fund for Nature

Contents

Authors Biography

Kamrul Hossain is an Associate Professor and Director of the Northern Institute for Environmental and Minority Law at the Arctic Centre of the University of Lapland. He has worked on a diverse range of Arctic issues mainly focusing on international environmental law and policy that applies to the Arctic as well as on human rights law, in particular, concerning the rights of the indigenous peoples, again with a focus on the Arctic.

Dele Raheem is a Senior Researcher at the Northern Institute for Environmental and Minority Law (NIEM) in the Arctic Centre, University of Lapland, Rovaniemi. His doctoral degree in Food Sciences was from Food and Environmental Sciences Department of the University of Helsinki, Finland. He has published in many peer-reviewed international journals in the fields of food microbiology, nutrition, food packaging and other related topics. His current research interest at NIEM is on crosscutting issues related to food and human security, especially in the Arctic-Barents region.

Shaun Cormier obtained his master's degree in law from the University of Lapland, Rovaniemi. He worked as a Researcher at the Northern Institute of Environmental and Minority Law (NIEM) at the Arctic Centre of the University of Lapland. His thesis was on "Enhancing Indigenous Food Security in the Arctic: Through Law, Policy, and the Arctic Council."

Chapter 1
Introduction

This book focuses on the Arctic-Barents Region, an area of northern Europe that comprises Norway, Sweden, Finland and Russia, with a population of 5.2 million, including indigenous peoples (Barentsinfo 2016a, b, c, d). The average population density in the region is 2.9 inhabitants per km², even though it includes sizeable cities, such as Murmansk and Archangelsk in Russia, Oulu in Finland and Umeå in Sweden (AMAP 2017a, b). The population of the region is composed of both indigenous and non-indigenous local inhabitants. The groups of indigenous peoples in the region include, for example, the Saami, Veps, Komi, Nenets, Pomors, Karelians. Some of these groups of indigenous peoples are transnational inhabitants. For example, the Saami live in four countries: Finland, Norway, Sweden and Russia's Kola Peninsula. The population of the region also transcends across the national and international boundaries to maintain their traditional livelihood activities. Some of these activities include herding, fishing, foraging and hunting, which are all practiced in cooperation with family members in group atmospheres and in their communities. Figure 1.1 is a map of the Barents region within the four countries (Norway, Sweden, Finland and Russia).

Food is regarded as an important element for both the local and indigenous populations of the region. Food is not just a commodity for physical consumption; it also offers cultural sustenance for the inhabitants of the region. Therefore, the traditional food habits have crucial value for the population of the region, since traditionally produced foods remain a vital component of healthy diets for the lives of the people living in this region. However, today food's access and availability is hampered by factors such as climate change and associated human activates—for example, increases in mining and oil and gas activities, tourism, shipping and forestry-related activities. The impacts from these factors have altered traditional food systems and activities, resulting in an influx of imported or store-bought "Western food" containing very little nutritional value due to being high in fat and sugar (Sheehy et al. 2015).

A better understanding of the food system in the Arctic-Barents region will help to shed light on how to improve the security and safety of foods in the region. Furthermore, the shift from traditional food has left northern communities with

© Springer International Publishing AG, part of Springer Nature 2018   1
K. Hossain et al., *Food Security Governance in the Arctic-Barents Region*,
https://doi.org/10.1007/978-3-319-75756-8_1

The Barents Region

70 N

Vadsø

Tromsø   Finnmark

Troms

Murmansk

Murmansk Oblast

Godø   Lapland   Naryan Mar

Nordland   Norrbotten   Rovaniemi   Nenets Autonomous Okrug

Luleå   Republic of Komi

Oulu Region

Oulu

Västerbotten   Kainuu Region   Arkhangelsk

Umeå   Kajaani

Republic of Karelia

Arkhangelsk Oblast

Sweden   North Karelia

Joensuu

Norway   Petrozavodsk

Finland   Syktyvkar

60 N

Russia

Fig. 1.1 Map of the Barents region highlighted in blue. (Adapted from: Barentsinfo.org. http://
www.barentsinfo.org/Barents-region. Accessed: August, 26, 2016)

health-related problems, such as increasing obesity and diabetes (Ford 2009). As a
consequence, the shift as such threatens not only the food security but also the
health security of individuals and communities. To counteract some of these consequences, there are a number of action plans, strategies and policies that have been
developed by institutions governing the region, such as the Barents-Euro Arctic
Council, the Arctic Council and the European Union (EU). In addition, there are
also national strategies in place. In this book, we bring the broader picture of food
security into the regional context. Therefore, we must first define food security for
use throughout this book.

A general definition of food security was endorsed by the Food and Agricultural
Organization (FAO) in 1996 (FAO 1996). The definition suggests that food security
exists when all people at all times have physical, social and economic access to sufficient, safe and nutritious food to meet their dietary needs and food preferences for
an active and healthy life (FAO 1996). While this definition offers a standard, it is
sometimes argued that it is flawed, since it relies heavily on the global economy and
does not necessarily capture the inherent issues of food security in the context of
helping active and communal voices to promote food preferences for an active and
healthy life (Schanbacher 2010). As described briefly above, given the presence of
distinct groups of peoples living in the Arctic-Barents region, their participation in
the practice of food preferences is important to establish so-called food sovereignty.
We integrated this aspect of sovereignty into our formulation of food security in the
Arctic-Barents region.

To some extent, the FAO definition of food security has overlooked the readily accessible traditional and local foods from hunting, fishing and other traditional forms of harvesting, even though they are part of the food preferences of the communities in the region. We invoke food security as integrated within the formulation of "food sovereignty." The World Forum on Food Sovereignty (WFFS 2001) defined the concept of food sovereignty as the ability and the right of people to define their own policies and strategies for the sustainable production, distribution and consumption of food that guarantee the right to food for the entire population. This definition also emphasizes the promotion of small and medium-sized production enterprises, where respect for inhabitants' own cultures is encouraged. It also ensures the diversity of peasant farming, fishing and indigenous forms of agricultural production, where the marketing and management of rural areas plays a fundamental role.

The shift toward food sovereignty encourages the production and harvest of traditional foods in a sustainable manner while discouraging food imports, which promotes innovative value addition approaches to traditional foods that are available within the locality (Pimbert 2009). As a result, we suggest that it is clearly relevant to the context of the Arctic-Barents region given the fact that the practice of food traditionally held by both indigenous and non-indigenous communities requires a stronger protection mechanism into which their voices are integrated. Only then can food security in the regional context be guaranteed.

By promoting and enhancing food security, it is likely that the closely overarching concept of human security can be further developed.

The concept of human security has been described as a way to protect the vital core of all human lives in ways that enhance their freedoms and fulfillment (CHS 2003). Food remains vital for human fulfillment, and it is directly related to many associated human rights and freedoms—for example, right to life, right to health, right to a healthy environment, right to water and right to culture. These rights are generally expressed in international human rights instruments and national constitutions of the countries around the world. The right to food remains a form of connection among people and their identity, culture and tradition. These elements are best exemplified in relation to food in the Arctic-Barents.

The use of traditional and local foods can help unite the Arctic-Barents communities and solidify the societal cohesion among their members through the promotion of new knowledge about food security in the regional context and across the borders. As mentioned, the Arctic-Barents region is being transformed rapidly as a result of stressors such as climate change and the new economic globalization, causing threats and challenges to food security; there is, however, very little documented research conducted on food security in the region in comparison to other areas of the Arctic, such as Canada and Alaska. Moreover, information on regulations and food policies in Russia is often limited in comparison to the other Barents countries, making comparison across the whole region a difficult task. We therefore shed light on how food security is impacted by various regional challenges and how the related governing institutions in place are offering incentives to promote greater food security. To promote and maintain a food-secure environment, it is important to look at the legal and policy tools that govern the region. While we look into a number of legal tools that offer a governance framework for the region having an effect on food security, we also analyze the role of relevant institutions and their strategies in this context. We examine the mechanisms associated with guidelines for issues such as food contamination, trade regulations, consumer safety, and public health. The purpose of our analysis of legal and policy mechanisms was to identify the existing gaps in knowledge, to understand the areas for improvement and to provide further suggestions and recommendations for enhancing food security in the Arctic-Barents region.

The methodology we employed in this exercise is as follows: we gathered previously published research studies related to food security in the region to review the existing food security situation in the region. Such a review provides us with important knowledge to produce an overall picture of the region as it relates to food. This was necessary for us to analyze the framework of governance prevailing in the region and how such framework is applied to promote food security. In addition to a literature-based analysis, we interviewed researchers in the fields of law, policy and the natural sciences who have had first-hand experience working with both the indigenous and non-indigenous peoples of the Arctic-Barents region. The results gathered from the interviews have been integrated in our analysis.

In this book, we highlight the major food security- and food safety-related challenges facing the Arctic-Barents region from a multidisciplinary perspective. The first chapter is an introduction to the book with a brief description of the Barents region and its relevance in the context of food security as it applies to the population of the region. In Chap. 2, we recognize the universality of food for all humans and elaborate why it should be a fundamental, guaranteed human right, with particular focus on the population of the Arctic-Barents region. In Chap. 3, we define food security as it falls within the framework of the concept of human security. We discuss the four major pillars of food security and integrate an approach to food sovereignty. Our effort is to show how the FAO definition of food security may not be completely applicable to the Arctic-Barents region given the reliance of the communities on traditional foods, which have an influence on the promotion of their culture, health and overall wellbeing. Chapter 4 follows this, where we discuss the traditional foods available in the Barents region and how local small food business operators add value to them. In Chap. 5, we raise the issues of food security among the Barents communities located in Finland, Sweden, Norway and Russia. In Chap. 6, we take a closer look at how the effects of climate change, new forms of human activities and the globalization of food sectors affect both food security and safety in the region. Concerning the issues of the governance of food security with reference to the Arctic-Barents region, in Chap. 7 we highlight the approach undertaken by the regulations governing the region as well as the role of the existing regional institutions that are relevant to promote food security. We therefore analyze the gaps in knowledge, and, based on this, we offer some recommendations in Chap. 8.

Chapter 2
Food Security: A Basic Need for Humans

2.1 Food and Human Security

The concept of security has long been displayed in its traditional form, as state-centered, which responds to threats with military force, as opposed to human security, which is addressed at the sub-state level targeting individuals and communities. The latter responds to threats arising out of multiples stressors, such as from both violent conflicts and issues such as environmental disasters, poverty, disease and human rights abuses (Owen 2004). The traditional security approach was predominant during the Cold War period, where people and countries were believed to be secure through military means by protecting the sovereignty of the states. However, the end of Cold War addressed a dramatic shift in the understanding of security by broadening and widening the very concept. The Human Development Report (HRD) of the United Nations Development Program (UNDP), endorsed in 1994, claimed that the concept of security had for too long been interpreted too narrowly—that is, the security of territory from external aggression or as a protection of national interests in foreign policy or as global security from the threat of nuclear holocaust (UNDP 1994). This has allowed key issues to fall through the cracks, as traditional security has failed at its primary objective of protecting individuals and communities (Liotta and Owen 2006). Heininen and Nicol (2007) redefined security by moving away from an exclusively state-centered and militarized geopolitical discourse to a more humanistic definition. This paradigm shift was best introduced and fostered by the 1994 United Nations HRD, which introduced the new security concept as "human security," allowing for the sustained protection of individuals and communities at the sub-national level. The Commission on Human Security, which was established in 2001 as an initiative undertaken by the UN Secretary General, explains that human security means protecting fundamental freedoms, freedoms that are the essence of life, being "free from want" and "free from fear" (CHS 2003). However, the concept requires more than just protecting people and their fundamental freedoms. To accomplish this goal, it is argued that

© Springer International Publishing AG, part of Springer Nature 2018
K. Hossain et al., *Food Security Governance in the Arctic-Barents Region*,
https://doi.org/10.1007/978-3-319-75756-8_2

the promotion of short-term protection from severe situations and threats as well as the promotion of the successful integration of political, social, environmental, economic, military and cultural systems and processes allow individuals and communities to foster the will and ability to sustain security and stability by themselves (Liotta and Owen 2006). Therefore, Liotta and Owen (2006), in their book, posed the question, "Why human security?" The answer to the question was provided in UN resolution 66/290 on human security, which highlights the centrality of human security as a universal framework that responds to a wide range of challenges and opportunities in the twenty-first century by seeking solutions that focus on efforts advancing the interconnected pillars of peace and security, development and human rights (UN 2013). The 1994 UN HRD elaborated the concept of human security with seven identified indicators of threats. They include threats from the potential lack of environmental, economic, health, community, political, personal and food security (UNDP 1994). The challenges to human security are interconnected, and they combine those to human rights, human development and peace and security. These issues have also been heavily addressed in the UN's millennium development goals (MDGs) (FAO 2016).

The application of human security and the principles that underpin it have proven to be essential in our combined efforts to advance the rights of people to live in freedom and dignity, free from poverty and despair and with an equal opportunity to enjoy all their rights and fully develop their human potential. Until we can ensure that people are safe, not just from interstate war and nuclear proliferation but also from preventable disease, starvation, civil conflict and terrorism, then we have failed in regard to the primary objective of security, which is to protect (Liotta and Owen 2006). One of the threats and insecurities that undermine communities and societies is food insecurity. Our focus region—the Arctic-Barents—provides crucial evidence where food security is argued to be threatened by multiple stressors resulting from an increase in various human activities. Also, in the regional context food is argued to promote a cultural bond among its inhabitants, where food connects families, groups and communities. Moreover, food security is linked to other dimensions of human security. Hence threats to food security, for example, contribute to threats to health security. The inter-linkage between food security and the other seven dimensions of human security—economic, food, health, environmental, personal, community and political—are shown in Table 2.1 (UN 2009; FAO 2016).

Henceforth, the concept of human security endorses an approach that is by definition people-centered, comprehensive, context-specific and prevention-oriented, addressing the plethora of risks and threats that endanger the resilience of communities (FAO 2016). Food security—as it connects the comprehensive human security approach from various dimensions, such as community culture, health and environment—in the Arctic-Barents region plays a vital role as it relates to the maintenance of a sustainable society.

The main challenge lies in creating a food system that will respond to impacts of climate change, which is the most alarming regional challenge resulting from human activities leading to the introduction of pollution, and thus endanger food security and safety in the region. The changes that are associated with livelihood

Table 2.1 The possible types of security and their human security threats[a]

Type of security	Examples of main threats
Economic security	Persistent poverty, unemployment
Food security	Hunger, famine
Health security	Deadly infectious diseases, unsafe food, malnutrition, lack of access to basic health care
Environmental security	Environmental degradation, resource depletion, natural disasters, pollution
Personal security	Physical violence, crime, terrorism, domestic violence, child labor
Community security	Inter-ethnic, religious and other identity-based tensions
Political security	Political repression, human rights abuses

[a]Based on the UNDP Human Development Report of 1994 and the Human Security Unit (Source: UN 2009)

patterns have affected how food is produced, processed, distributed and consumed. In many regions of the world, the stressors resulting from volatile food prices, erratic weather, natural hazards and competition over resources are increasingly leaving millions of vulnerable populations in insecure conditions (FAO 2016). These changes lead to challenges concerning the formulation of policies to address food security in the Arctic-Barents context, especially in light of stressors such as climate change, the rising cost of living and changes in food sharing networks. Food insecurity, as said, also includes other detrimental effects—adverse effects on human health—which have multiplying negative effects on human security (UN 2012).

Striving for food and nutritional security and empowering people to build resilience can help the most vulnerable to face the risks and overcome the resulting shocks that can threaten their security (FAO 2016). Since the human security approach is predicated on being people-centered and building the capacity of individuals, the approach provides key tools for building resilience in regard to food security and nutrition (FAO 2016). The relationship between human security and food security is also located in the connection to their prevention-oriented approach. The promotion of food security surely provides a conflict mitigation/prevention tool in many ways, such as a sustainable community built around the peaceful practice of societal norms traditionally rooted in a given society, for which food serves as an important indicator. Hence, food security is crucial in developing overall human wellbeing, which ultimately strengthens human security.

This has been recognized in the Arctic-Barents region. The Nordic Forum for Security Cooperation report in 2014 showed that the formats of both the Arctic cooperation and Barents cooperation promote security in different ways through their inclusive, broad and bottom-up approaches (NFSC 2014). Thus, the main areas of priority are to help promote economic, cultural and social development in the North through building knowledge and human capital. This is essential for further developing the Arctic-Barents in a way that allows all aspects of human security to be considered. The Barents region has experienced and will continue to experience the use of its natural resources. The Finnish Committee on European

Security, for example, emphasized the need to extend knowledge about the interaction between traditional and new industries pertaining to environmental, socio-economic and cultural aspects (NFSC 2014) to further promote a number of human security aspects. The report emphasized the indigenous cultures and traditional languages as well as the practices held by the indigenous communities, such as hunting, fishing and reindeer herding activities (NFSC 2014). To foster the human security approach when dealing with food security issues, it is important to develop a bottom-up and all-inclusive approach. Therefore, the human security approach requires that the people of this region have a voice in future decision-making processes and that their suggestions be heard so that economic development can continue without harming the human security and human rights of individuals and communities. It should be noted that human survival is not connected only to physical sustenance; intellectual, cultural and spiritual survival require the preservation of cultural heritage. The contribution of the cultural field to this vision can be realized through continuous and interdisciplinary efforts to extend culture as an integral element of society (OPM 2014). The Arctic-Barents region as such has a distinct cultural heritage, with the presence of traditional communities whose survival as being free from want, fear and indignity requires the preservation of their unique culture, and the protection of food practice offers a defense from many of the aspects of human security threats (Larsen and Fondahl 2015).

2.2 Food as Human Rights

The human right to food and association with food can mean a number of things, but in most instances, it deals with the right to food itself and the right to adequate food. Moreover, it also ensures that individuals have the means to the access and availability of food such that people do not go hungry or starve. The basic human rights associated with food are in place to protect individuals so that they have the ability to utilize food to maintain their health and lives. These rights not only promote food security and the four pillars of food security—availability, access, utilization and food systems' stabilization—but they also further enhance the human security concept.

Human rights are universal, often expressed through and guaranteed by law, in the form of treaties, customary international law, general principles and other sources of international law. International human rights law lays down the obligations of governments to act in certain ways, or to refrain from certain acts, to promote and protect human rights and the fundamental freedoms of individuals or groups (OHCHR 2016). Furthermore, human rights are rights that are inherent to all human beings, whatever their nationality, place of residence, sex, ethnic origin, color, religion, language or any other status (Brown 2016). In the Arctic-Barents region, both indigenous and non-indigenous peoples are fully entitled to universal human rights as guaranteed by international, regional and national human rights instruments. The FAO determined that the relationship between human rights and food security should be an idea for the full realization of the human right to adequate food as a fundamental human right and one that leaves no one behind (FAO 2016). In addition to the right to adequate food, every man, woman and child, alone or in community with others, has the right to physical and economic access at all times to adequate food or the means for its procurement (UDHR 1999).

Ever since the right to food and the right to adequate food were mentioned in the 1948 UN Declaration of Human Rights, they have since gained ground in other national and international instruments. The "right to food" is also conveyed through two norms, the fundamental right of everyone to be free from hunger and the right to adequate food, which are substantially different from one another. The freedom from hunger is the only one that qualifies as a "fundamental" or "absolute" standard by the International Covenant on Economic, Social and Cultural Rights (ICESCR) (Bultrini 2009). This right is an inclusive one and therefore not strictly a right to a minimum ration of calories or nutrients but to all the nutritional needs of a person to live a healthy and active life and the means to access or attain them (OHCHR 2016).

Bultrini (2009), however, explained that the right to adequate food is much broader, as it implies the existence of an economic, political and social environment that will allow people to achieve food security by their own means. The right to food, as part of an adequate standard of living and a fundamental right to be free from hunger, acknowledges that there are many related factors, such as poverty and health care. In addition, the right to food has three separate elements, including availability, accessibility and adequacy (Haugen 2012). *Availability* in this sense means that food should be available either through the production of food, by culti-vating land or animal husbandry, or through other ways of obtaining food, such as fishing, hunting or gathering and also from its sale at markets and shops. *Accessibility* means that the economic and physical access to food must be guaranteed. In this case, "economic" implies that food must be affordable without compromising other basic needs, while physical access implies that food is accessible to all (including those physically vulnerable and children). The *adequacy* of food must satisfy the dietary needs, taking into account factors such as the individual's age, health and occupation, for example. The food should be safe from adverse substances and be culturally acceptable.

However, there are a variety of factors that must be fulfilled to protect food as a human right, very similar to that of food security. It is also significant to note that the right to food is not the same as a right to be fed. Some suggest this as a frequent misconception, where the assumption is that governments are required to hand out free food to anyone in need. The Office of the High Commissioner for Human Rights makes it clear that individuals are expected to meet their own needs, through their own efforts and using their own resources, and to be able to do this, a person must live in conditions that allow him or her either to produce food or to buy it. To produce his or her own food, a person needs land, seeds, water and other resources, and to buy it, one needs money and access to the market (OHCHR 2016).

Human rights associated with food are vital for the further promotion of human security; these rights further ensure that food security is promoted and maintained among individuals and groups. Viewing human rights as the baseline protection for food and basic needs for further food security calls for the baseline protection for human security and the further promotion of human security. Therefore, to consider how this can be better promoted, there are a number of examples of human rights norms that promote food security, which are discussed in the next paragraphs.

The right to food and the right to adequate food are displayed in many national and international documents protecting the human rights of individuals around the world. In addition to the 1948 Universal Declaration of Human Rights (UDHR), the International Covenant on Economic, Social and Cultural Rights (ICESCR), the International Covenant on Civil and Political Rights (ICCPR), the Convention on the Elimination of All Forms of Discrimination against Women (1979) and the Convention on the Rights of Persons with Disabilities (2006) offer norms in relation to the right to food. The most articulated clause determining the right to food is probably Article 11 of the ICESCR (1966), which reads as follows:

1. The States Parties to the present Covenant recognize the right of everyone to an adequate standard of living for himself and his family, including adequate food, clothing and housing, and to the continuous improvement of living conditions. The States Parties will take appropriate steps to ensure the realization of this right, recognizing to this effect the essential importance of international co-operation based on free consent.
2. The States Parties to the present Covenant, recognizing the fundamental right of everyone to be free from hunger, shall take, individually and through international co-operation, the measures, including specific programs, which are needed:

   a. To improve methods of production, conservation and distribution of food by making full use of technical and scientific knowledge, by disseminating knowledge of the principles of nutrition and by developing or reforming agrarian systems in such a way as to achieve the most efficient development and utilization of natural resources;
   b. Taking into account the problems of both food-importing and food-exporting countries, to ensure an equitable distribution of world food supplies in relation to need.

The obligations enumerated in this Article are legally binding upon ratifying states, meaning that the right to food must be upheld and promoted. In the Barents region, these international agreements have been signed and ratified by all countries. To complement these legally binding international agreements, there are, in addition to the obligation articulated in the human rights instrument, soft-law documents that further protect the right to food that are not binding nor obligatory for states but rather voluntary and serve as a form of guidance for the implementation of the right to food, such as policy papers produced on food security (ICC 2012).

Another example is the "Right to Food Guidelines," adopted by the Council of the FAO (FAO 2005). The guidelines are designed to assist states and non-state actors in how they can successfully implement their existing obligations in relation to food security. According to Bultrini (2009), the "Right to Food Guidelines" can help governments design appropriate policies, strategies and legislation. Although voluntary, the guidelines can have a significant influence on state policies (Bultrini 2009). In addition, the Human Rights Council within the UN has created a special position called the Special Rapporteur on the Right to Food. Bultrini explains that the importance of the Special Rapporteur is to clarify the contents of the right to food by explicating its meaning to governments regarding their obligations in respect to this right (Bultrini 2009). This is fundamental in assisting countries to better comprehend the outcomes and benefits of adopting necessary measures on the right to food.

The Committee on Economic Social and Cultural Rights (CESCR) is an authoritative body created under the ICESCR, which has adopted General Comment 12 (GC12) on the right to adequate food. The GC12 explains that every state is obliged to ensure access to the minimum amount of essential food, which is sufficient, nutritionally adequate and safe. Food should also be available and accessible (Lundqvist et al. 2015). As we have already discussed, availability and accessibility are both crucial for food security; the GC12 proposes three levels of obligation for states, which are to respect, to protect and to fulfil the obligation of the right to adequate food and food security (CESR 1999). Lastly, the United Nations Declaration on the Rights of Indigenous Peoples (UNDRIP) is also an important document, where food is placed in specific relation to indigenous peoples. The UNDRIP, although not legally binding, is important for focusing specifically on indigenous peoples' rights. For example, its Article 1 states that indigenous peoples have the right to the full enjoyment, as a collective and as individuals, of all human rights and fundamental freedoms as recognized in the Charter of the United Nations (UNDRIP 2007). This is significant for those human rights discussed above, such as the right to adequate food in relation to these peoples. Furthermore, it goes a step beyond in recognizing not only individual but also collective rights to such human rights.

The right to food surrounds the notion of individual food security. This poses a problem, as the right to food must not be limited to strictly the individual but rather viewed as part of the collectivities. An individual right refers to those rights enjoyed by an individual person, separate from others, while collective rights refers to those rights enjoyed by the group as a whole. In current human rights treaties, the individualistic approach to rights and rights-holders is portrayed the most (UNRIC 2016). This narrow focus in current treaties can present a problem for indigenous peoples, who usually self-identify as an individual and then connect themselves to the larger community to which they belong (Hossain 2015a, b, c). This is most relevant in the Barents region, where we see that the connection between community and food is demonstrated throughout the process of hunting, gathering, processing and even in consuming the food, in particular by groups of indigenous communities. Even though indigenous peoples respect their guaranteed individual rights under the current human rights mechanisms, they often advocate for further protection of collective rights. Such is the case under the right to food, first enjoyed by individuals, but indigenous peoples also frequently exercise this right collectively (Knuth 2009). By requiring additional collective rights under the right to food, indigenous peoples are said to be continually reshaping the boundaries of human rights and its relationship to food.

The group component is implied in the UDHR in that the right to food is a right that can be also enjoyed collectively, which is relevant for all indigenous peoples. While the obligation under the UDHR does not provide any binding force, food as part of cultural rights can also be demonstrated from Article 27 of the ICCPR, which explains that in those states in which ethnic, religious or linguistic minorities exist, "persons belonging to such minorities shall not be denied the right, in community with the other members of their group, to enjoy their own culture, to profess and practice their own religion, or to use their own language" (ICCPR 1966). Indigenous peoples base their reasoning on collective rights in connection to their language, religion and culture, and the food practices constitute an important aspect of culture. Collective rights in relation to food activities are important for those groups that have traditionally been involved in the collecting of food, fishing and herding. In the Barents region, the collective component of rights is of importance as it relates to food security, in particular for indigenous peoples.

Although the right to food and the right to adequate food are important for the entire population in a given society, there are, however, many other connected human rights obligations, and the promotion of these rights is also beneficial to promote food security. The Office of the High Commissioner for Human Rights has highlighted a number of rights, such as the right to health and food, the right to life and the right to water (OHCHR 2016). Adequate nutrition can be used as a component for both food and health. For example, if a pregnant or breastfeeding woman is denied access to nutritious food, she and her baby can be malnourished even if she receives pre- and post-natal care (OHCHR 2016). The right to life is threatened when people are not able to feed themselves with adequate and nutritious food, resulting in the risk of death by starvation, malnutrition or resulting illnesses. The right to water is also important in that the right to food cannot be fully realized when people lack access to safe drinking water for personal and domestic uses (OHCHR 2016).

Some have indicated that the right to food should also include the right to water, as one cannot have food without the access to and availability of water. Food production and preparation both require access to water, so further guidance is needed on how to interpret the interconnection between the human right to food and that to water. For example, in Russia, fishing is one of the main economic activities for the Saami, because they usually live on the coastal areas of the Kola Peninsula. They are also protected under law, because Russian legislation grants the right for indigenous peoples to use the water for the purpose of traditional fishing. This includes not only a priority right to choose hunting and fishing areas but also the exclusive right to hunt and fish in such areas by following certain time schedules (NFSP 2014). In fact, this is for all groups of indigenous peoples in Russia, as these groups need to have access to water and its resources to sustain their livelihoods (NFSP 2014). If the "right to adequate housing" is lacking—in the sense that a house lacks basic amenities, such as for cooking or storing food—the "right to adequate food" of its residents may be undermined. Also, when the cost of housing is too high, people may have to cut down on their food expenses. This has been documented in the Alaskan Arctic, although it is also relevant to the Arctic-Barents and Canadian Arctic, where climate change is influencing the underground storage of food and making it prone to pathogens and bacteria due to the warmer temperatures (Brubaker et al. 2009). Temperature increases and changes in rainfall patterns have a profound impact on the persistence and patterns of occurrences of bacteria, viruses, parasites and fungi and the patterns of their corresponding foodborne diseases (Tirado et al. 2010). Furthermore, when housing prices are too high in the Arctic, individuals feel the need to save money on food, resulting in the choice of less nutritious and cheaper store-bought food. This trend could also be taking place in the Barents region.

The *right to education* and the *right to information* are also important, as information is crucial to make the right choices in regard to the right to food (La Rue and Elham 2015). It enables individuals to know about food and nutrition, markets and the allocation of resources. It also strengthens people's participation in decision-making and consumers' choices (Graham 2015). Although traditional foods may not be packed and labeled with nutritional information as with store-bought foods, it is nevertheless important that consumers have good knowledge about their food choices, food ingredients and other nutritional education. Thus, access to information to be able to identify chemicals in food, and their possible effects on the body, must be linked to the right to information, which also connects with the right to food.

Concerning obligations on providing adequate protection of the right to food, all four countries in the Barents region are parties to the most significant and legally binding documents, including the international bill of rights—the combination of the UDHR, ICCPR and ICESCR. In addition, these countries are also parties to, for example, the Convention on the Elimination of All Forms of Discrimination against Women (1979) and the Convention on the Rights of Persons with Disabilities (2006), which contain clauses on the right to food.

However, the protection regime for the right to food is found to be rather weak in the Arctic-Barents region, since the Optional Protocol to the ICESCR is not applicable to the region. The Protocol offers a clear guarantee that states that fail to comply with the obligation can be held accountable. Article 2 of the Protocol reads as follows:

> … communications may be submitted by or on behalf of individuals or groups of individuals, under the jurisdiction of a State Party, claiming to be victims of a violation of any of the economic, social and cultural rights set forth in the Covenant by that State Party. Where a communication is submitted on behalf of individuals or groups of individuals, this shall be with their consent unless the author can justify acting on their behalf without such consent. (ICESCR 2009)

This Article is important for both individuals and groups in the Barents region in regard to bringing forth complaints that affect their economic, social and cultural rights. Since none of the countries in the Barents region is a party to this Optional Protocol, the guarantee of the right to food is based on a vague legal formulation in terms of complying with the human rights obligation under ICESCR. However, it is important to mention that the "right to food" is also an inalienable right as part of states' human rights obligation under their national constitutions.

2.3   Conclusion

As much as the right to food is connected to the right to life, the assurance of food security for citizens is simply a fundamental obligation. According to Knuth (2009), Finland recognizes the implicit right to food as a part of the broader human rights framework, whereas Sweden, Norway and Russia do not offer any direct reference to the "right to food" in their constitutions. However, as discussed in this chapter, states' obligation under international law regulating the general human rights framework offers the promotion of food security all across the globe. The Arctic-Barents region, consisting of the countries with the most advanced democratic practices in the globe, in particular the Nordic ones, are in the forefront of protecting human rights for all. Elsewhere in this book, we analyze the performance of these countries in regard to the governance of food security to promote the right to food as it pertains to the population of the region. It is, however, important to consider the food security situation from the regional context as it relates to Arctic-Barents region, which we discuss in further detail in the next chapter.

Chapter 3
General Background: Food Security in the Arctic-Barents Region

3.1 Food Security: A Conceptual Framework

Food has increasingly become a topic of discussion and debate worldwide with the rapid growth of the global population and concerns over the ability to sufficiently feed everyone. Food security evolved as a concept after the former American President Franklin D. Roosevelt stated that food was "the first want of man." He later developed the term to include, "the freedom from want and the freedom from fear" (Akram-Lodhi 2009). According to Roosevelt, food security indicates the availability of an adequate and suitable supply of food to secure human needs. Moreover, he viewed "secure" as referring to the accessibility of food, "adequate" as referring to the quantitative sufficiency of the food supply and "suitable" as referring to the nutrient content of the food supply (CFS 2012).

Thus in the context of food "freedom from want" offers a secure, adequate and suitable supply of food for every man, woman and child. At the first World Conference held in Rome in 1974, a declaration was adopted that highlighted as its aspiration that no child would go to bed hungry in subsequent 10 years (WFC 1974). Twenty years later, the 1994 UNDP HDR report referred to freedom from want as the freedom from chronic threats such as hunger, disease and natural disasters. Alternatively, freedom from fear is an approach strongly held by Canada as referring to violent conflicts arising out of, for example, inequity, the incapacity of states to ensure security, ethnic conflict and discrimination. Broadly, the definition of human security, as indicated earlier, is largely placed within the parameters of violent threats against the individual, which can stem from a vast array of issues, including the drug trade, landmines, ethnic discord, state failure and trafficking in small arms (McRae and Hubert 2001). However, "fear," as it relates to remaining free from violent conflicts, also indicates that an inadequate supply of food unquestionably results in threats of violent conflict. Thus, ensuring food security from a fear perspective refers to the notion that one should not have to worry about having adequate food, or in other words, one should not have to be stressed about where his

© Springer International Publishing AG, part of Springer Nature 2018　　　　15
K. Hossain et al., *Food Security Governance in the Arctic-Barents Region*,
https://doi.org/10.1007/978-3-319-75756-8_3

next meal will come from. There have been various attempts to define food security at different fora, but nowadays, the definition widely recognized is that which is endorsed in the final report from the World Food Summit in 1996, published by the FAO (1996). The foundation of the FAO (1996) food security definition consists of four pillars: availability, accessibility, utilization and food systems' stability. This concept primarily concerns peoples whose environment has witnessed changes in these four pillars. The Arctic-Barents region, as referred to earlier, is drastically witnessing changes in these four pillars due to the ongoing transformations facing the region. Food, being a crucial human need for survival, remains important among both the indigenous and non-indigenous populations of the region, through its use, consumption and sharing.

The four main pillars (availability, accessibility, utilization and stability) are defined here to convey the concept of food security. The *availability* of food is determined by the physical quantities of food that are produced, stored, processed, distributed and exchanged (FAO 2008). Food availability looks not only at the traditional and local food perspectives but includes imported foods as well. *Accessibility*, according to the FAO, is a measure of the ability to secure entitlements, which are those set of resources (legal, political, economic and social) that an individual requires to obtain access to food. In terms of accessibility to traditional foods, the issues surrounding their safety must also be discussed. Food *utilization* refers to the appropriate nutritional content of the food and the ability of the body to use it effectively—in other words, the safety and social value of food. Lastly, food systems' *stability* is about the removal of uncertainty and the promotion of an effective, constant and balanced supply determined by the temporal availability of and access to food (FAO 2008). As long as these four main pillars are in place, a population or individual is considered to be food secure.

Although these four pillars provide a conceptual framework, which is now widely recognized in the understanding of food security, some researchers (e.g., Windfuhr and Jonsén 2005; Holt-Giménez and Shattuck 2011; Nilsson and Evengård 2015) have suggested that it is inadequate by failing to represent all interests involved in the securitization of food. In particular, groups of indigenous peoples claim that the current concept of food security is insufficient, since the understanding oftentimes relies on the assessment of monetary access to market food without focusing on the need for the primary consumption of traditional foods harvested from the land. Therefore, many have argued that the definition of food security for indigenous people should include the assessment of traditional food intake and the stability of access to it (Egelund et al. 2013).

3.2 Food System and Its Impacts on Food Security

The food system is better conceptualized from a multi-stakeholder approach that takes into account its complexity to ensure that the four main pillars of food security are guaranteed. The food system is a combination of the activities taking place from

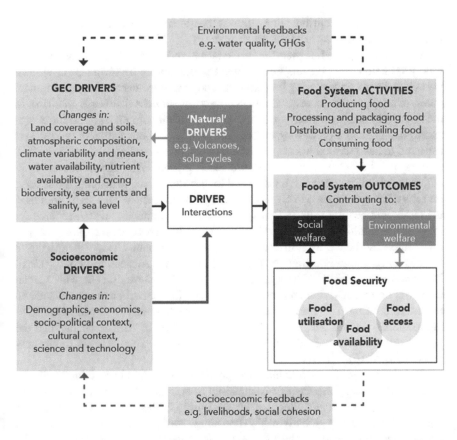

Fig. 3.1 The interactions and complexity of the food system. © Ingram J., Ericksen P. and Liverman D. (Eds.) 2010. Food Security and Global Environmental Change, London, Earthscan

the farm to the plate and their outcomes. It also includes the drivers such as global environmental change and socio-economic drivers that affect these activities and outcomes. The outcomes entail various effects on food security, including social and environmental impacts, health consequences, employment and ethical issues.

Figure 3.1 below shows the interacting components, feedbacks and drivers. These interactions within the food system will have a significant impact on the food security in any society. By analyzing the activities that take place within the food system, and how they are governed, we can better understand their resulting outcomes.

Concerning environmental footprint, the food system contributes to approximately 20–30% of global human-made greenhouse gases (GHGs) although there is vast inherent uncertainty in these estimates. The use of fertilizers, pesticides, manure, farming and land-use change and livestock are huge contributors of human-made GHG emissions (Vermeulen et al. 2012). In addition, later stages in the food system, such as packaging, retail, transport, processing, food preparation and waste disposal, combined contribute to 5–10% of the global GHGs although their importance and likely impacts are forecast to grow (FCRN 2016).

A systemic approach is therefore necessary to reduce GHG emissions linked to food production and consumption to help the agriculture sector adapt to climate change while not endangering food security and to achieve sustainable development goals, in particular in the restoration of ecosystems services (Muller et al. 2016). In the Arctic-Barents, a region where the impacts of climate change are most pronounced, an efficient processing of local and traditional foods with reduced waste is expected to have limited impact on the environment. According to some scenarios, meeting the 1.5 °C limit would require much greater GHG reductions by 2030 (Climate Analytics 2016). Agriculture and food systems need to adapt to the adverse impacts of climate change to ensure resilient food production. This starts with having fertile and healthy soils and recognizing their importance as a key resource for long-term agricultural production (Muller et al. 2016). There are measures in place, for example, at the EU level to address the impact of climate change on agriculture and food processing. For example, the new EU Common Agricultural Policy (CAP) is aligned to the UN's 2030 Agenda for Sustainable Development with special emphasis on incentivizing and rewarding the tangible, environmental and societal outputs of farming. This measure aims to provide high-quality food while contributing to the EU's goals for rural viability, the mitigation of climate change and the promotion of environmental conditions that will help to keep farmers in business (Falkenberg 2016).

Organic agriculture also holds good potential to contribute to climate change mitigation and many aspects of sustainability in agriculture, such as improved soil quality and biodiversity (Muller et al. 2016). During the last decade, the demand for organic food has risen globally, which encourages organic farming. In 2015, about 6% of agricultural areas in the EU were under organic management, while in the Barents region, there are differences between the countries. Northern Finland takes the lead in regard to organic management in this region (Muller et al. 2016). Organic agriculture has a strong focus on enhancing and maintaining the fertility and quality of soils, and a number of its core practices support that goal. Practices such as cover crops, mulching and intercropping protect soils against erosion from both run-off water and wind with less reliance on fertilizers. A gradual increase in organic agriculture to 50% by 2030 will reduce nitrogen levels and compensate for soil carbon sequestration emissions by 10–15% of the total agricultural emissions of the EU (EC 2016). This is in 2030, it is expected that the mitigation potential of organic agriculture within the EU will reach about 30% (Muller et al. 2016).

Human health has been shown to benefit from an increase in organic production in the EU, and the market for organic food is growing. One of the most comprehensive meta-analyses carried out to date, including 343 peer-reviewed publications (of which approximately 70% were studies carried out in Europe), indicated that organic food differs from conventional food in the concentration of antioxidants, pesticide residues and cadmium (Cd) (Barański et al. 2014).

It is important to define certain terms surrounding foods and their context in the Arctic-Barents. Local food, as defined by Ludger Muller-Wille (2001), refers to any food produced or harvested in the environment of a community, and this is regardless of whether it was produced for subsistence or commercial purposes (Duhaime and Bernard 2001).

There can be different foods under this definition, including traditional food, which entails a group that belongs to a defined geographical space, and it is part of a culture that implies the cooperation of the individuals operating in that territory (Bertozzi 1998). Local food is also associated to food miles; it refers to the distance food is transported from the time of its production until it reaches the consumer. Food miles are used when assessing the environmental impact of food, including the impact on global warming (Engelhaupt 2008). For traditional food, a food product must be linked to a territory, and it must also be part of a set of traditions, which will necessarily ensure its continuity as a traditional food for certain people. Traditional food is vitally important in the Arctic-Barents region, particularly concerning indigenous peoples, as these foods are said to ensure their identity through the promotion of culture and traditions that the communities carry over generations. Additionally, "white food," store food and/or store-bought food are frequently used terms designated to those foods coming from outside of the region. In a similar context, imported food refers to any product manufactured outside of the region whose initial acquisition requires that there is a monetary exchange or an operation that involves a business transaction within the community or through a southern wholesaler (Duhaime and Bernard 2001).

3.3 The Arctic-Barents Region and the Importance of Food Security

The Arctic is a vast landscape, often portrayed and perceived as a desert full of ice and snow, but this is not the full picture. The Arctic-Barents is a unique environment that is properly used and has been sustainably maintained with an abundance of food. The Barents region, for example, has a varied geography, where some say it is characterized by its remoteness, harsh climate and varied landscapes, with the Scandinavian mountain chains in the west, the Arctic tundra in the Kola Peninsula and the Nenets Area and Novaja Zemlja in the east (BEAC 2016a, b, c). The geographical area spans a distance of 1.75 million km², the midnight sun is up almost 24 h a day from May to July, and the phenomenon of the northern lights in the autumn and winter months is common. The Barents region is surrounded by boreal forest and thousands of lakes and mountains, which provide the region with an abundance of available natural food. In fact, this region is said to contain more forests, fish, minerals, oil and gas than any other region in Europe, making it a rich and plentiful region of natural resources.

The Barents region consists of 13 counties and many different ethnic groups in four countries (Välkky et al. 2008). According to Välkky and colleagues, there are at least 15 different languages spoken in the region. Out of the 5.5 million people

Table 3.1 The 13 counties of the Barents region[a]

Country	County	County center
Norway		
	County of Nordland	Bodø
	County of Troms	Tromsø
	County of Finnmark	Vadsø
Sweden		
	County of Västerbotten	Umeå
	County of Norrbotten	Luleå
Finland		
	Province of Lapland	Rovaniemi
	Province of Oulu	Oulu
	Province of Kainuu	Kajaani
Russia		
	Arkhangelsk region	Arkhangelsk
	Republic of Komi	Komi
	Murmansk region	Murmansk
	Republic of Karelia	Petrozavodsk
	Nenets Autonomous Area	Naryan-Mar

[a]Adapted from Välkky et al. (2008)

that live in the region, 1.6 million live in the Nordic countries and 3.9 million in Northwest Russia. In 2005, the largest city in the region was Murmansk, with a population size of 380,000, followed by Arkhangelsk, with 356,000 inhabitants. The largest Nordic city is Oulu (Finland), with approximately 129,000 inhabitants, followed by Umeå (Sweden), with a population of 106,000.

Table 3.1 below shows the 13 counties in the four countries that make up the Barents region.

One of the major characteristics of the Barents region is its sparse population. The territories are vast, which makes the distances long, creating challenging conditions for developing infrastructure in the region. In the Nordic countries, both road and railway systems are well developed, whereas in Northwest Russia the road network is less developed. Travel and long distance transportation are mainly carried out through railroads. Generally, in the Barents region, there is a lack of east–west connections, as most of the connections go from north to south (Välkky et al. 2008). There was a rail connection to Russia from Kemijärvi in Finland that existed until about 50 years ago. Sweden and Norway have only one rail link in the Barents region, between Narvik and Lulea, and none between Finland and Sweden. In regard to the water transport network, the region has a number of major ports that form important links. Murmansk is northern Russia's largest port that operates throughout the year, and its importance for commercial shipping is on the rise. Other important Russian ports include Severodvinsk, Arkhangelsk and Kandalaksha; and Narvik in Norway as well as Kemi in Finland and Lulea in Sweden on the Gulf of Bothnia are also important ports in the Barents region (EC 2000).

Historically, fishing, hunting, agriculture and herding were all common in the Arctic-Barents region. The populations indigenous to the area have practiced these activities for many generations, and the non-indigenous who eventually migrated to the region learned this specific way of life from engaging with the locals (Kelman and Næss 2013). These activities then developed over time to incorporate new aspects; for example, it is said that reindeer were first a wild species hunted by indigenous peoples but have been gradually domesticated with ownership rights (Castro et al. 2016). These engagements between indigenous and non-indigenous peoples opened up and led to an increase in interregional trade during the Middle Ages. During this period, indigenous peoples played an important intermediary role in the early transboundary trade relations, as public markets only first appeared in the 1500s (BEAC 2016a, b, c). At the same time, traders and merchants living in the European part of northern Russia, in the present day Barents region, would attend fairs and trade fish, fur and game for commodities such as hemp, bread and clothes (Elenius et al. 2015). After the Middle Ages, the Pomor trade, an intensive trade system that existed between northern Norway and the Murmansk–Arkhangelsk region of Russia, was established as a common tradition. The Pomor trade greatly enhanced the economic and cultural development of the region. The trade between Norway and Russia grew from five ships at the end of the seventeenth century to 400 ships in the eighteenth century. In that period, a new pidgin language, "Russenorsk," emerged, which was a mix of Russian and Norwegian (Jahr 1996; BEAC 2016a, b, c), as a consequence of the promotion of trade between the regions. Russian merchant ships from the White Sea area used to come loaded with goods that were in short of supply in Northern Norway, such as grain and flour, canvas and linen, hemp and rope, iron goods and tar. They exchanged these for fish, in which the Russians were not self-sufficient (BEAC 2016a, b, c). The early markets and trade in the region allowed for increased interaction and communication among people in the area with the sale and trade of food between societies, cultures, and communities in the Barents region (Jahr 1996).

In 2000, a field study on food security was conducted by Ludger Muller-Wille and co-workers in the village of Kylä (in the Savukoski municipality of northern Finland). Their study found the main sources of income for villagers to be reindeer herding, fishing, forestry, berry picking, employment, unemployment benefits and pensions (Muller-Wille et al. 2008). At the time, Kylä consisted of both Finns and Saami populations, who had lived there for generations. Within the village and municipality, people would sell reindeer meat and fish; however, the local exchange or sharing of food such as reindeer meat, fish, berries and potatoes was very rare, as it mostly occurred within families and between younger and older generations. One could also use meat and fish as a means of payment services, such as the leasing of hayfields or snow removal. This barter and trade type system has historically been quite common for connecting people and communities in this region. Although technology has advanced, and markets come in different forms nowadays, the traditional ways of trading and communicating around food has remained the same. Kylä did not have a grocery store, but this was not a problem, as people who have cars often go to the nearby municipality center or they make use of the mobile store that comes from a nearby village to deliver food products to certain households twice a week.

The continuous access to resources is of principal importance to ensure access to food. It is particularly important for those areas where rural communities spread out in sparsely populated regions, such as the Arctic-Barents region. The geographic closeness to these resources guaranteed their continuous use, especially in terms of traditional food and food security. The fieldwork in Kylä referred to herein demonstrates an example of how small communities in the Barents region use food as a means of connecting people and communities, perhaps not to the same degree as it was once, but it is still quite important.

3.3.1   Food and the Environment in the Barents Region

The relationship between food security and the prevailing climatic conditions in the North remain poorly understood and under-examined in the existing body of knowledge. Also, there is little information available on the inter-relationships between climate change and food security as they relate to human health being jeopardized due to regional transformations (Ford 2009). With the projected accelerated changes as a result of the effects of climate change, there is a need to gain a deeper understanding of these relationships (Ford 2009; Tong et al. 2010).

The preservation of the natural environment is important to the Barents region. In fact, it was the initial starting point for the cooperation between Norway, Sweden, Finland and Russia in the Barents. This further led to the Arctic-wide environmental cooperation initiated through the Rovaniemi Declaration, the so-called Finnish Initiative of 1991, which also included the other circumpolar Arctic nations (Canada, Denmark, Iceland and the U.S.) in addition to those forming the Barents region. These nations created the Arctic Environmental Protection Strategy (AEPS) with a view to promoting safeguards for the protection of Arctic's pristine environment (Russell 2008). At a review on the assessment of the Arctic Council's work upon its tenth anniversary, questions on the determination of the type of treaty arrangements and the provisions that are most appropriate for the Arctic were raised (Koivurova and VanderZwaag 2007). Some suggested that improving the living conditions of those peoples inhabiting the Arctic and the Barents was the primary aim of the cooperation, as living conditions are closely associated with the environment (Rafaelsen 2010; Sellheim 2017).

According to Muller-Wille et al., the food security concept is closely associated with the historical development and gradual expansion of environmentalist ideas and practices in the Western world (Muller-Wille et al. 2008). They also highlighted that the environmentalist discourse initially started as being primarily about the natural sciences but has now dynamically extended to include the human dimension (Muller-Wille et al. 2008). *Barentsinfo*, a portal for regional information, adds that changes in the environmental and social conditions are interdependent, whereby environmental conditions and trends affect human health and quality of life. In addition, there is a need to review social conditions and outcomes when designing and implementing environmental management activities and policies (Barents Info 2016a, b, c, d).

This discourse between food and the environment goes further than the theoretical and political perspectives, as there is a true human dimension to this concept when food security is considered. The relationship that exists between food and the environment had been established long before any theoretical or political discussions in the region. Eventually, the Kirkenes Declaration reached in 1993 laid a foundation for a strong economic and social development in the region with emphasis on an active and sustainable management of the nature and resources (BEAC 2016a, b, c). The Arctic Human Development Report (AHDR) also measures social indicators about human development on a long-term basis. For example, a sustainable Arctic-Barents region requires access to safe food and water that are free from contaminants (Rover and Ridder-Strolis 2014).

Most of the European Arctic is biologically richer and more productive than other Arctic areas because of the warming effects of sea currents and air masses (BEAC 2016a, b, c). At the same time, the natural environment is unique in its biodiversity and is an important part of the Earth's ecosystem, providing living resources that form the basis for human settlement in the region. The taiga in the Barents region is low in species richness, but many species are found year-round or at least part of the year. These include reindeer, moose, red deer, roe deer, mountain hare, beavers, squirrels and voles. In the region's taiga, predators such as the Eurasian lynx, stoat, European otter, wolverine, gray wolf, red fox and brown bear are found (AMAP 2017a, b). The region's tundra hosts resident mammals such as the Arctic hare, Arctic fox, ptarmigan, lemming and reindeer (AMAP 2017a, b).

This unique biodiversity is attracting contamination as climate change impact proceeds. The Arctic Monitoring and Assessment Program (AMAP) recently highlighted that there is a general lack of information concerning the extent to which emerging chemicals may be taken up by and accumulated in Arctic fauna and indigenous Arctic peoples, whose diets depend heavily on local wildlife (AMAP 2017a, b). AMAP's work has subsequently played an important role in bringing about the Stockholm Convention on Persistent Organic Pollutants, which was signed in 2001 and took effect in 2004 (AMAP 2014). At that time, contaminants such as polychlorinated biphenyls (PCBs) and dichlorodiphenyltrichloroethane (DDT), which had been in use since the 1940s and 1950s, were carried by winds and water currents from industrialized regions in the South to the Arctic, where they ended up inside the bodies of seals and polar bears found the Arctic-Barents region (Arctic Now 2017). When humans eat these marine mammals, they are affected, too; for example, there were high levels of PCBs found in the bodies of Inuit women of child-bearing age, which were associated with infertility and cancers (Arctic Now 2017).

The Arctic-Barents region, when compared to the rest of the globe, is expected to be wetter and warmer as a result of climate change. The effects of global warming on biodiversity has been an interesting research topic, especially in the context of this particular region. In a Nordic Council of Ministers' technical report by Hof et al. (2015) on the future of biodiversity in the Barents region, the impacts of climate change on different terrestrial species in the region were highlighted. The report claimed that terrestrial vegetation is likely to be dominated by needle-leaved forest, whilst grasslands will become very rare, and species richness may increase or decrease in the Barents region in the future. Dispersal ability will affect species' richness—that is, if species are able to disperse and fully utilize their future climatic niches, then the species' richness will increase (Hof et al. 2015). If, on the other hand, species are not able to disperse beyond their current climatic niche, but are able to maintain in the areas they occupy at present, their richness is expected to decrease (Hof et al. 2015). Therefore, it has been suggested that there is a need to protect areas in the coastal regions of Fennoscandia and in the southwestern parts of Northwest Russia, since these areas are going to be climatologically increasingly suitable for a large number of species. The future of the biodiversity in the Barents region may also be undermined in these areas since they have the potential for greatly increasing human activity having an effect on the geo-political scenario at large and socio-economic consequences for the region's population (AMAP 2017a, b).

The populations in the Barents region have utilized its natural resources found in lakes, fields, mountains and forests to establish long-term settlement. The stable human settlement in the region is thus dependent on the protection of the natural environment and its ecosystem. The natural environment and the communities inhabiting the region are connected in various ways. For example, the holding and the use traditional ecological knowledge (TEK) are integral to these communities. According to the World Intellectual Property Organization (WIPO), TEK is the knowledge, know-how, skills and practices that are developed, sustained and passed on from generation to generation within a community, often forming part of its cultural or spiritual identity (WIPO 2000). In 2000, WIPO members established an Intergovernmental Committee on Intellectual Property and Genetic Resources, Traditional Knowledge and Folklore (IGC); and in 2009 an international legal instrument (or instruments) was developed to give traditional knowledge, genetic resources and traditional cultural expressions (folklore) effective protection (WIPO 2009). The communities in the Arctic-Barents have been closely tied to TEK for generations, in particular concerning the promotion of their traditional food system.

Since TEK provides incentives for the sustainable management of eco-systems to promote food security, such knowledge has long been utilized to create a bridge between the natural environment and human needs. For example, reindeer herding—"*siidas*"—in the Barents region has traditionally utilized such knowledge for generations. The term *siidas* refers to an ancient Saami community system within a designated area, but it can also be defined as a working partnership in which the members have individual rights to resources but help each other with the management of the herds or when hunting and fishing (ICR 2017). The *siida* could consist of several families and their herds; the system emphasizes kinship organizations and the year-round tending of reindeer, and the *siida* societies have a long-established connection in the Barents region through their need for survival and sustainable use of the food and the environment (Sara 2009).

In these societies food security is bound to the principles and practices of local land use and the management of natural resources. Reindeer herding is typically a highly space-extensive means of livelihood, requiring vast pastoral range and long migrating routes (Beach 1981). Nowadays, both Saami and Finnish reindeer herding vary from free grazing to large fenced systems or systems of intensive herding by kin or village communities with a controlled circulation of pastures (Heikkinen and Sarkki 2010). Human activities and developments in the region can alter both the natural and infrastructural environments with their impact on the sources of traditional food supply. Today, modern agriculture and infrastructure have caused excessive pressure on the same land areas where reindeer used to graze freely. In the region, the traditional lands are increasingly utilized for the development of, for example, roads, rails, dams, mines and tourist areas. The infrastructural changes affect subsistence livelihood practices. Reindeer herding is said to be the most important producer of food items in this area. Since food and the environment are intertwined, and mutually reinforce each other, it is vitally important to have proper environmental management processes. For example, the outcomes from the developments referred to herein impact the food security and living conditions of the communities directly unless a proper environmental management process is in place. The results of environmental management policies clearly include references to the assurance of local food resources and thus their security (Muller-Wille et al. 2008).

3.3.2   Traditional Food, Health and Physical Survival

In an era when "security" has acquired a new gravitas, the term is no less politically loaded for representatives of the traditional communities inhabiting the Arctic-Barents region. For example, the Inuit community and their ancestors managed to feed themselves from the land and sea where Europeans perished. Yet a recent survey showed that nearly 70% of Inuit preschoolers surveyed live in a household deemed food insecure (Egelund et al. 2010). In this survey, primary caregivers reported a range of scenarios from worrying about running out of money to buy food for preschoolers to not eating for a whole day. Such insights have prompted formal calls for programs that can improve the price of food to make it affordable and to make both market and traditional foods available. These efforts have taken the form of subsidies for airfreight costs of fresh foods along with educational activities toward a greater public awareness of food preparation and healthy eating practices.

Despite the efforts to promote healthy eating, there are concerns over water-, air- and food-borne diseases that are on the increase in the Arctic (SliCA 2015). Since food production accounts for 70% of all human water use, in the processing of these foods, quality aspects are paramount, which usually start from the growing conditions of the ingredients, including water. The quality of water provided by different municipalities differs greatly between the Nordic countries and Russia. For example, the quality of tap or well water is not monitored regularly in the same way in all the municipalities of the Barents region. The metal levels in household water in six cities of the Murmansk region (Nikel, Zapolyarny, Olenegorsk, Montchegorsk, Apatity and Kirovsk) showed that some cities lack sanitary protection zones for water sources, that most cities require preliminary water processing and that the method of water disinfection involves only chlorination (Dudarev et al. 2015). High levels of aluminum in water were found in Kirovsk, and high levels of nickel in the water of Zapolarny and Nikel cities were found. Water taken from the Petchenga region's springs demonstrated relatively low levels of metals, except for strontium and barium (Dudarev et al. 2015).

It was observed that the Arctic populations consume traditional and local foods that may be exposed to pollutants released into the environment of remote and local sources (Donaldson et al. 2010). For example, the Arkhangelsk region is an industrially developed area, with large pulp and paper industries. Rautio et al. (2017) reported that emissions from industrial facilities in the region have a high content of sulfur dioxide (50%) as well as various kinds of dust (16.5%), carbon monoxide (10.6%), hydrocarbons (12.6%) and nitrogen oxides (9.35%). It was also observed in 2006 that the cities of the Arkhangelsk region are witnessing an increase in the level of air pollution with nitrogen dioxide and particulate matter that can compromise the quality and safety of these foods (Bogdanov et al. 2011).

Thus, human health and wellbeing both in rural and urban communities are subject to potential threats given that societal and environmental transformation in the regional setting is taking place. The threats as such also lead to human development challenges as a result of demographic changes, such as population aging, migration and urbanization. The increasing economic interest in the Arctic also brings new population groups, who work for short periods in the industries—for example mining or sailing sectors, and leave the region eventually. This applies to the Barents region, as it changes the population dynamics, and there is a need to follow the parameter trends of wellbeing and an environmentally sound and healthy lifestyle. It was observed that there has been a rise in chronic diseases because of this transformation of lifestyle as it relates to the practice of food consumption, which is partly due to a shift to a more Western diet with more junk food, less physical activity and high levels of smoking and alcohol and drug abuse (Rautio et al. 2013).

The health aspects of food in the Arctic-Barents is usually associated with traditional foods. Such foods, often also referred to as local foods, remain vital for the populations belonging to both indigenous and non-indigenous communities for their physical survival and health benefits. Before market food was available in stores, people had to physically go out onto the land to obtain their food, primarily through means of hunting, fishing and gathering (Nuttall et al. 2005). The physical aspect of this work was usually strenuous but could also be beneficial to the health of the individual. Physical activity was largely more a part of the daily life in the 1930s–1950s than now (Nilsson et al. 2011). In the present-day context, it has been found that there are no major differences in reported physical activity between the indigenous and non-indigenous populations except for reindeer-herders. The present reindeer herders execute physically strenuous work more often than non-reindeer herding Saami and non-Saami in the region (Nilsson et al. 2011). The benefits of such strenuous physical work are best explained through the convincing evidence of a protective effect of physical activity on colorectal cancer and the evidence for a likely similar effect in regards to breast and endometrium cancer (WCRF 2007). There is also evidence for the beneficial effects of physical activity related to cardio-vascular diseases (CVDs), both for general physical activity levels (PALs) and specific activities such as walking (Nilsson 2012). The efforts and additional energy spent on hunting and gathering food, as well as the related cleaning, cutting, picking and preserving, in terms of health are far more beneficial to the human body than driving to the grocery store and collecting food. In addition to positive physical benefits, there are many positive health aspects related to consuming these foods.

Despite the well-documented influx of imported foods, traditional foods remain an important part of the diets among the northern communities because of their high source of nutrients. Traditional foods are seen as having far greater health benefits and as far superior to store-bought or imported foods (Nuttall et al. 2005). This was also confirmed through a study conducted at the University of McGill, Canada, which analyzed both traditional and imported foods in 43 Canadian communities. The study found that on days when people ate both traditional and market foods, their diet was better than when only eating market food. For example, when traditional foods were consumed, the benefits were found to include fewer calories, which is helpful for weight control, and the consumption of more lean meats from game animals and fish leads to less saturated fat, which is better for the heart. By consuming such foods, one obtains more minerals and vitamins, such as iron—good for the muscles and blood, zinc—good for wound healing and fighting infection, vitamin A—good for vision and fighting disease and calcium—good for strong bones and teeth (CINE 2016). The study also concluded that the sharing and consumption of these foods also strengthened cultural capacity and wellbeing in the communities. In another study of autopsied Inuit men in Greenland, Dewailly and co-workers found almost no indication of prostate cancer, a finding they suggested could be related to the intake of wild food rich in omega-3 polyunsaturated fatty acids and selenium (Dewailly et al. 2003). The consumption of traditional foods is still appreciated in many Arctic communities. A 2006 federally administered survey on 6300 Inuit children and adults across the Arctic revealed that in 65% of homes, "country food" such as seal, caribou, whale, duck, fish and berries accounted for at least half of the food consumed in the household (Tait 2006).

In another study conducted in 2006, it demonstrates the relationship between traditional foods and health under the framework of the International Barents Secretariat project "Revealing the hidden diabetes mellitus in Lovozero district of Murmansk Oblast," 4359 residents (2736 rural and 1623 urban) of Kola Lapland in Murmansk Oblast were interviewed and had their blood glucose levels analyzed. Participants included 694 Saami residents, 910 Komi residents and 80 Nenets residents (AMAP 2015). The results showed that the risk of type 2 diabetes (overweight/obesity, enhanced blood pressure, sedentary lifestyle, malnutrition, alcohol abuse) was three- to sevenfold lower in indigenous residents than non-indigenous residents (AMAP 2015). Signs of diabetes were noticeably absent among the Saami people of the remote villages, while elevated blood glucose levels were found mainly in large settlements. Indigenous residents in remote villages demonstrated a minimal risk for diabetes mellitus, and this may be related to their traditional diet based on local foods, a physically active lifestyle and minimal consumption of high carbohydrate foods (AMAP 2015). Therefore, traditional foods are vital not only for protecting but also for promoting health and wellbeing among Arctic-Barents populations. The protective effects of these bio-components can be expected to diminish as traditional food becomes a less prominent portion of the diet.

Shifting away from traditional foods, relying more on imported foods and engaging less in physical activity has led to a higher incidence of cancer in the circumpolar region. An international circumpolar review of cancer among the Inuit populations of Alaska, Canada and Greenland during the period 1989–2003 showed high risks for lung, nasopharynx, colorectal and salivary gland cancer among the Inuit in comparison with non-Inuit groups as a result of changing lifestyles, dietary transition, decreased physical activity levels and changes in socio-environmental conditions (Kelly et al. 2008). Although these studies were based in the Canadian Arctic, many of the same foods found in the Barents region have similar nutritional value and benefits. Although the knowledge about the health and living conditions among the Saami is poor, the epidemiology of cancer in the Saami population has been investigated in several studies (Sjölander 2011). The major reason for studying the epidemiology of cancer is that large areas within Sápmi (the cultural region traditionally inhabited by the Saami people in Fennoscandia) were contaminated by nuclear fallout as a result of atmospheric nuclear weapons testing on the island of Novaya Zemlya in the 1950s and 1960s (Hassler et al. 2008). Also, the Ukrainian Chernobyl nuclear reactor accident in 1986 makes the Sápmi region an interesting area for fieldwork on epidemiological studies (Hassler et al. 2008). The observations made during the Saami cancer research add to the impression that the low risk of developing prostate cancer is a common trait among the native people of the Arctic circumpolar region (Mahoney and Michalek 1991). These traits are part of the genetic makeup that has been passed on from one generation to the next; the interactive role of diet and genes is an important topic.

The genetic structure of the Saami population makes it suitable for studies on how genetic and environmental factors influence the development of common diseases. Differences in incidences of heart disease were investigated in studies that reflect the ongoing transition from a traditional to a more Westernized lifestyle (Brustad et al. 2008; Nilsen et al. 1999; Håglin 1991). One of these studies confirmed that the nutrient density in the Saami and lumberjack diets was well above recommended levels for most nutrients except for folate and fiber (Håglin 1999). Lumberjacks are loggers of the earliest times, they had no modern logging equipment, they worked in lumber camps and often lived a migratory life, following timber harvesting jobs (Rohe 1986). Håglins's (1999) study on the traditional Saami diet was compiled from interviews with old Saami people living today and from information available from the prevailing body of literature. Genetic factors have been suggested as contributing to ethnic differences in prostate cancer, and lifestyle factors—such as diet and physical activity—also dominate the discussion on why the Saami are less likely to develop cancer of the prostate (Haldorsen and

Tynes 2005; Soininen et al. 2002; Hassler et al. 2001). In a study among the Swedish Saami, an analysis of prostate cancer in relation to lifestyle and genetic heritage indicated that the main reason for the lower risk of developing prostate cancer points to lifestyle rather than to genetic factors (Hassler et al. 2001).

There is an ambiguity in terms of traditional Saami food in relation to lifestyle regarding the risk for gastrointestinal cancer. Their diet consists of foods that can increase the risk of cancer (e.g., smoked and salted meat and fish and a low intake of fresh fruits and vegetables) and those that can decrease its risk (e.g., reindeer meat; wild fish that are rich in selenium, omega-3 acids and vitamin A; and a low intake of dairy products) (Ross et al. 2006). The extent to which diet influences genetic modifications is an important topic of study that can shed light on this ambiguity. Dietary compounds are known to regulate epigenetic modifications that can provide significant health benefits and prevent various pathological processes involved in the development of cancer and other life-threatening diseases (Vahid et al. 2015).

There have been many studies on the genetic origin of the Saami but only a few studies investigating health-related genes. Most of these studies have focused on alleles that are linked to easily measurable biomarkers, such as the plasma lipid profile. One of these is the APOE, a class of apolipoprotein E that is the principal cholesterol carrier in the brain (Puglielli et al. 2003). This protein combines with fats in the body to form lipoprotein molecules. The APOE protein plays a key role in plasma lipoprotein metabolism and in lipid transport within tissues (Davignon et al. 1988). These lipoproteins are responsible for packaging cholesterol and other fats and carrying them through the bloodstream (Eichner et al. 2002). Diet can affect the level of cholesterol, and maintaining normal levels of cholesterol is essential for the prevention of disorders that affect the heart and blood vessels (cardiovascular diseases), including heart attack and stroke (Verschuren et al. 1995).

The APOE gene encodes three alleles or alternative forms—APOE 2, APOE 3 and APOE 4. APOE 2 is associated with low levels of total plasma cholesterol, low density lipoprotein (LDL) cholesterol and apolipoprotein B. APOE 4 shows the opposite pattern and is associated with their increased levels (Sing and Orr 1976). APOE 4 is a major risk factor for susceptibility to coronary heart disease and Alzheimer's disease, particularly when combined with a Western diet. APOE 3 genotype is considered to be neutral. Among the Saami and several other indigenous populations, the frequency of APOE 4 is high (31% in the Saami), while among other populations with a long history of agriculture, such as the Greeks, the frequency is low (5.2%) (Davignon et al. 1988). APOE 4 is also relatively common (17.4–20.8%) in Swedes, Danes and Finns. The APOE genotype has been shown to influence plasma antioxidant status, with increased antioxidant levels for APOE 2 (Ortega et al. 2005). The frequency of this allele among the Saami is 5% (Comas et al. 1999).

The high frequency of the APOE 4 allele in the Saami contributes to their susceptibility to coronary heart disease, given exposure to the appropriate environmental factors. The lower coronary heart disease risk observed in earlier studies is likely to be the result of their lifestyle; so with the transition to a more Westernized lifestyle, the high frequency of the APOE 4 contributes to even higher coronary heart disease risk than is the case for other populations in the area. The gene encoding apolipoprotein

A4 has also been studied among the Saami (whose parents are both Saami) and the Finns (whose parents are both Finns) living in northern Finland. There are two common alleles at this locus, the frequencies of which differ between the Saami and the Finns (10.6% and 5.6%, respectively) (Lehtinen et al. 1998). In the Saami, the heterozygote for these two alleles had a higher high-density lipoprotein (HDL) than APOE 4.1 homozygotes (Lehtinen et al. 1998). Slightly higher levels of total cholesterol, LDL, HDL and triglyceride levels have also been found in the Saami than in the Finns. However, recent studies on the plasma lipid profile of the Saami have not found any differences from geographically matched controls (Edin-Liljegren et al. 2004). There is a need for more detailed analyses of the genotype/phenotype combinations to identify specific risk groups within the Saami population (Ross et al. 2006).

The inheritance of adult-type hypolactasia and the occurrence of hypolactasia in different countries around the world as well as the amounts in their populations are suggested to be linked to genetic and dietary factors. Lactose tolerance in the Swedish Saami varies between 40% and 75% for different subpopulations (Sahi 1994), which is much lower than in the general Swedish population (91%). For example, the ability to digest lactose as an adult has been associated with two different mutations located upstream of the lactase gene LCT (Enattah et al. 2002). The haplotype with these two mutations was shown to have been under positive selection 5000–10,000 years ago in many European citizens, consistent with its selective advantage in dairy farming cultures (Bersaglieri et al. 2004). The Saami have been involved in reindeer herding over the last 1000 years and used reindeer milk, which is very low in lactose (2.4%), on a limited basis until the 1920s. It is suggested that the consumption of reindeer milk for generations could have shaped the high frequency of lactose tolerance in the Saami (Bersaglieri et al. 2004). However, with a relatively short exposure to dairy products with a high lactase level from admixture with the European farming population led to a strong genetic drift (Kozlov and Lisitsyn 1997). This shows how exposure to a new diet may have negative nutritional effects on genetic makeup.

In the future, genetics will play vital roles in relation to diet and health with the emergence of nutrigenomics. Nutrigenomics investigates the use of molecular tools to search for, access and understand the various responses that are obtained from the diets of individuals and of population groups (Pavlids et al. 2015). The relevance of nutrigenomics in the health sector is due to the massive body of work on the conclusion of the Human Genome Project (1990–2003), the role of diet on epigenetic modifications and the need for personalized medicine.

3.4 Food and Its Relations to Cultural Wellbeing

Food is not only a commodity for physical consumption for survival; it also brings cultural sustenance, in particular for traditional communities. For Arctic-Barents communities, traditional foods have long served as a crucial element for health as well as for spiritual and cultural wellbeing (Muller-Wille et al. 2008). The cultural diversity in the Barents region is displayed through the populous indigenous and non-indigenous peoples that have resided in the region who have used food as a way of connecting to their diversity and culture. This complies with the famous saying "you are what you eat," and it truly applies in this context. As highlighted by Nordstrom and colleagues, one of the basic foundations of a society rests on its eating habits; hence there is no culture without food (Nordstrom et al. 2013). Cultural wellbeing is about having the freedom to practice one's own culture and to belong to a cultural group as promoted by, for example, the Canadian Research Institute for the Advancement of Women (CRIAW 2016).

Activities such as hunting, herding, fishing and gathering are based on continuing social relationships between people, animals and the environment (Nuttall et al. 1992). These activities remain significant for maintaining social relationships and cultural identity within indigenous societies. They define a sense of family community and reinforce and celebrate the relationships between them and their surrounding natural environment upon which they depend (Nuttall 1992; Callaway 1995). The cultural aspect and relationships with animals and the natural environment extend much deeper in rich mythologies. The vivid oral histories, festivals and animal ceremonialism also illustrate the social, economic and spiritual relationships that indigenous peoples have with the Arctic environment (Nuttall 1992; Callaway 1995).

Furthermore, the natural environment has a spiritual essence as well as cultural and economic value, and the land and water that surround the communities are regarded as cultural commodities. For indigenous peoples, there are many features in their landscape regarded as sacred, especially along migration routes, where animals reveal themselves to hunters in dreams or where people encounter animal spirits while travelling (Brody 1983). Once again, reindeer herding in the Barents region provides a perfect illustration of the cultures and traditions of the peoples that are thousands of years old. The consumption of food from animals is, therefore, fundamentally important for personal and cultural wellbeing. Across the Arctic as well as in the Barents, the sharing and distribution of meat and fish is central to daily social lives, which express and sustain social relationships (Nuttall 1992). This is evident throughout the whole process of gathering, hunting and consuming the food (Solstad 2012). Yet, despite the importance for social identity and cultural life, the primary need for and the use of animals are based purely on a need for survival (Nuttall 1992).

As the culture itself transforms, the culture around the food is also being rapidly transformed. Today, access to new technology, such as the use of the internet and satellite television, has given the northern inhabitants an unprecedented awareness of the wider world, while a money-based economy has given them goods manufactured by that world. Snowmobiles and powerboats have largely replaced dogsleds and kayaks just as high-powered rifles long ago replaced hunting spears. Where people once harvested their own food from the surrounding land or sea, today many buy groceries in stores, including processed and packaged products that would have been unknown a few decades ago (Lougheed 2010).

Comparing food practices between the North and South can even be culturally different. Jorge Jordana found that the range of traditional foods consumed in the North was narrow and without much elaboration, with the objective of eating primarily for nutrition (Jordana 2000). On the other hand, he claimed that the food of southern countries is consumed in variety and not just because of the favorable weather conditions, given that many of the basic ingredients are imported and prepared in sophisticated ways, clearly seeking the pleasure inherent within food and drink (Jordana 2000). Food consumption habits can be influenced by a series of factors—such as the fulfilment of a basic need, the desire for pleasure, the structure of supply, income levels, ostentation or the fact of belonging to a particular culture with its religious or moral characteristics—and there may be a major difference between the North and the South in this regard (Jordana 2000).

Food choices go hand-in-hand with changes in lifestyle. Among the traditional communities in the Barents region, while lifestyle transforms the adaptation to new developments, environmental sustainability is highlighted as the most significant aspect linking human settlements with the securities of food, water and energy (AHDR II 2015). Food forms part of who we are and can become, and it also ties us to our families to create an identity (Almerico 2014). It is established that human psychological needs can intertwine with social factors when foods are used more for the meaning they represent than for the nourishment they offer or provide (Brown 2011; Almerico 2014).

Chapter 4
What Types of Foods Are Available in the Arctic-Barents Region?

4.1 Imported Versus Traditional Foods

When Finland and Sweden joined the EU in 1995, they became the EU's northern-most countries in the Barents region. The Barents region is of special significance to the EU not only for its economic potential and great environmental value but also as its only direct border with the Russian Federation. On a general level, there is a significant trade relationship between the EU and Russia. The EU is Russia's largest trading partner, accounting for 40% of Russia's exports and 38% of Russia's imports (EC 2000). Trade between the EU and Russia has decreased since 2012—by about 44% between 2012 and 2016 and from €339 billion in 2012 to €191 billion in 2016 (EC 2017).

There is no doubt that imported foods are entering the Arctic communities from other parts of the world at a faster rate than ever before due to globalization. It is difficult to determine the exact figures, especially in the Barents region, since the region is less researched than other areas of the Arctic such as Canada and Alaska. This lack of data on food consumption patterns in the Barents region was reported more than 15 years ago, and the situation is relatively the same even today (Duhaime and Bernard 2001; Duhaime and Godmaire 2001). Duhaime et al. (1998) made it clear that although so-called traditional food remains a central element in the discussions surrounding aboriginal culture, imported food comprises the greater share of consumed food. These foods, the so-called store-bought foods, such as hamburg-ers, hot dogs and pizza, offered by the rapidly expanding chains of fast food restau-rants in even the smallest settlements have become one of the main sources of food and nutrition security for the local population (Duhaime and Bernard 2001). This trend is not only in one place; it is happening in all across the circumpolar north at different rates. These changes are of serious concern to the health and wellbeing of the population living there and consuming these foods.

At the same time, it is important to understand that this is not a recent phenom-enon; the process has been occurring over the past decades and centuries. As a

© Springer International Publishing AG, part of Springer Nature 2018 33
K. Hossain et al., *Food Security Governance in the Arctic-Barents Region*,
https://doi.org/10.1007/978-3-319-75756-8_4

yardstick for the Arctic-Barents region, we traced the origin of imported or exogenous food to the Saami food system (Lansman 1999). Historically, it is understood
that the Saami food system did not exist in isolation due to some form of interaction
with other communities. External relations with surrounding peoples to the south
and through trade in markets have resulted in the steadily increasing introduction of
exogenous food (imported food) items that, over centuries, have altered the dietary
habits and conditions of northern peoples considerably. These exogenous foods are
those originating from distant resources, whose exploitation and utilization is under
the control of others who have developed global markets for their distribution
(Ruong 1969).

Exogenous food items of the past whose consumption and nutritional value may
be doubtful today, but whose recreational and cultural value were highly placed
within the Saami household in the past are, in succession, tobacco, beer, coffee, tea
and sugar (Lansman 1999). It is believed that these food items were introduced from
the South and readily accepted and integrated into the region sometime in the seventeenth century. The changes in food consumption were substantially documented
by Ludger Muller-Wille at the turn of the twenty-first century. Recent changes in the
pattern of food consumption started to occur in the 1930s, before World War
II. During this period, a study conducted by Israel Ruong, a Saami, on the changing
foundations of the Saami food economy, which had experienced a mounting degree
of dependency on external food commodities, referred to this process and situation
as "distant consumption" (Ruong 1969; Muller-Wille 2001). The historical records
indicate that for centuries, food security was provided and guaranteed for the Saami
by a large array of living resources accessible endogenously from the fauna, flora
and water available in the boreal and subarctic environments (Itkonen 1948; Muller-
Wille 2001).

In the 1990s, local Saami in retrospect agreed that living off the land was feasible
in the 1950s and 1960s with little reliance on external food items. However, close to
the twentieth century they saw a rise in exogenous goods and foods as well as external institutional influences—that is, central governments or special groups such as
sport fishers were gradually displacing or diverting the endogenous food resources,
thus drastically limiting their availability to local residents. Although many of these
endogenous food resources were used traditionally by the Saami, they are still part
of their diet today but to a much lesser degree. One suggestion for this change was
the rapid establishment of wage employment and the use of cash, which have dra-
matically altered the economic structure of the Arctic-Barents region over recent
decades. Consequently, this has gradually led to the disappearance of subsistence
hunting and fishing activities. In moving forward, the traditional and indigenous
foods need to be processed and marketed in the same way as exogenous food items
to have a chance to compete in the global market. Many have recognized the need
for such market and the need to sell such foods internationally, emphasizing their
health and cultural benefits (Gellynck et al. 2012). Although we highlighted the
Saami as an example here in this section, changes in food consumption have
occurred across the whole of the Barents region in a similar way.

The exact statistical figures for imported food and traditional food consumption for the whole Barents region are difficult to determine. However, some numbers from Sweden might paint a general picture. It is said that only 3% of Swedes hunt, but nearly 40% fish at least once a year, and 23% regularly participate in fishing (Romild et al. 2011; Arlinghaus et al. 2015). It is difficult to determine if these numbers are high in comparison to other countries in the region or how they might differ in Northern Sweden. The current game meat harvest in Sweden is estimated at 16 million kilograms annually, of which 11 million kilograms come from moose. In comparison, although the values are different, in Norway an estimated 35,000 elk, 40,000 roe deer, 20,000 red deer and 15,000 reindeer are hunted every year (Jarratt 2014). They also say that tens of millions of kilograms of berries and very significant quantities of mushrooms are harvested each year in Sweden (Adams 2017). Some hunters have indicated that in the northern and rural regions, around 30 years ago almost all protein was wild-hunted or fished. Nowadays in northern Sweden, only a little more than 20% of their food comes from regional production, and all other foods are imported in refrigerated trucks. In the early 1990s, nearly 112,000 reindeer were slaughtered in Sweden on an annual basis, and in recent years only half this number have been slaughtered annually (FCES 2014). The economic growth of small-scale farming in the Barents region is hampered, as it does not seem possible to combine any production of food in the Arctic-Barents region with competition from the more efficient agriculture sector on the European continent. The Barents data pool shows a declining trend in the number of cattle in the Barents region from the period between 1999 and 2012, where the number went down by almost one-third in the Tromsø county of Norway (Staalesen 2015).

When there is a lack of traditional foods, market and store-bought foods may provide food security in terms of availability. From that perspective, imported foods, despite their poor quality in regard to fulfilling the dietary needs of the region's population, ensure accessibility regardless of the question of affordability. The statistics presented above do not paint a perfect picture of the rate or amount of imported foods entering into the Barents region though it is clear that the trend shows a decrease in traditional foods and an increase in market foods. Regardless of the statistics, they are not the determining factor here in this book, as our focus is primarily on a general overview of the legal and policy aspects toward the promotion of better food security. It is still important, however, to refer to the food consumption trends that are taking place in the Barents region and the whole of the Arctic. Table 4.1 is an indication of the cereal grains (wheat, barley, oats and rye) that were produced in the Barents region from 2008 to 2012. In Finnish Lapland, the figures were for winter and spring rye. There was no cereal production in Norwegian Troms and Finnmark, likewise in Russian Nenets and Murmansk.

Table 4.2 shows the amount of potatoes produced in the Barents region from 2008 to 2012. It was possible to grow potatoes in all the regions, especially in Finnish Northern Ostrobothnia.

Table 4.1 Cereal grain production (100 tons) in the Barents area of Finland, Sweden, Norway and Russia (2008–2012)

	2008	2009	2010	2011	2012
Finland					
Northern Ostrobothnia	214.5	291.7	239.3	259.3	230.0
Kainuu	10.9	12.3	10.5	10.6	8.4
Lapland	4.6	4.0	2.1	1.2	2.4
Sweden[a]					
Norrbotten	9.9	9.2	9.6	4.2	
Vasterbotten	21.7	20.9	19.4	20.5	
Norway					
Nordland	0.8	0.5	0.0	0.6	0.7
Troms	0.0	0.0	0.0	0.0	0.0
Finnmark	0.0	0.0	0.0	0.0	0.0
Russia					
Republic of Karelia	0.13	0.05	0.08	0.02	0.01
Republic of Komi	0.79	0.02	0.04	0.0	0.0
Arkhangelsk Oblast	0.73	0.3	0.0	0.0	0.0
Nenets AO[a]	0.0	0.0	0.0	0.0	0.0
Murmansk Oblast	0.0	0.0	0.0	0.0	0.0

[a]There were no data for Sweden or Nenets AO in 2012 (Source: (1) OSF Tike, crop production statistics [www.maataloustilastot.fi]; (2) Swedish Board of Agriculture [www.jordbruksverket.se]; (3) Norwegian Felleskjøp statistics; (4) Fedstat [www.fedstat.ru])

Community remoteness and northern latitude often restrict access to fresh and nutritious market foods. Generally, food costs in the Arctic are high—for example, comprising 23–43% of household income in the Russian Arctic (Dudarev et al. 2013). Due to climate change many wildlife species consumed as country foods have disappeared. The shortening of the snow-covered period in winter influences human travel and transportation. Recent studies indicate elevated rates of household food insecurity in many places in the Arctic. For example, in Nunavut, Canada, nearly 70% of the Inuit preschoolers and in Chukotka, Russia, 45% of the indigenous people have been found hungry during recent years (Egelund et al. 2010).

Traditional foods are characterized by a link to a certain territory and were defined by Bertozzi (1998) as part of a culture and imply the cooperation of the individuals belonging to that territory. There are many foods in the circumpolar north and the Barents region that match Bertozzi's definition of traditional foods in terms of identity, tradition and culture. Many of these traditional foods are associated with indigenous peoples and their culture; however, non-indigenous people can also have these same connections (Figs. 4.1 and 4.2).

Fishing (i.e., subsistence fisheries) by indigenous peoples has been historically extensive throughout the Arctic. The available freshwater in the region as well as diadromous fishes—those of the species that regularly migrate between fresh and marine waters—are of particular importance to humans both inside the Arctic-Barents and elsewhere.

Table 4.2 Potato production (100 tons) in the Barents area of Finland, Sweden, Norway and Russia (2008–2012)

	2008	2009	2010	2011	2012
Finland					
Northern Ostrobothnia	62.9	106.7	111.6	114.9	89.6
Kainuu	2.5	2.2	1.6	1.5	1.0
Lapland	2.8	3.4	2.6	2.8	2.0
Sweden[a]					
Norrbotten	11.2	10.1	9.5	7.6	
Vasterbotten	7.2	6.7	6.4	7.1	
Norway					
Nordland	4.2	1.6	3.2	3.2	1.8
Troms	4.1	6.4	2.2	6.2	2.1
Finnmark	0.2	0.2	0.1	0.2	0.1
Russia					
Republic of Karelia	16.0	13.73	16.87	16.08	12.22
Republic of Komi	14.68	8.8	4.24	8.19	10.9
Arkhangelsk Oblast	41.44	24.02	24.78	26.37	27.58
Nenets AO[a]	0.0	0.0	0.0	0.0	0.0
Murmansk Oblast	4.38	4.17	5.67	6.18	4.23

[a]There were no data for Sweden or Nenets AO in 2012 (Source: (1) OSF Tike, Crop production statistics [www.maataloustilastot.fi]; (2) Swedish Board of Agriculture [www.jordbruksverket.se]; (3) Norwegian Felleskjøp statistics; (4) Fedstat [www.fedstat.ru])

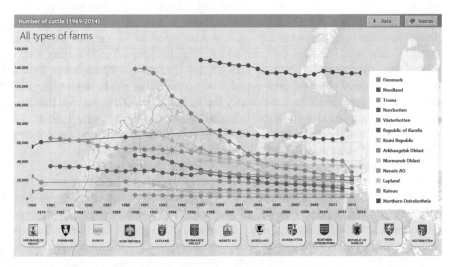

Fig. 4.1 The number of cattle in the Barents area of Finland, Sweden, Norway and Russia (1969–2014). (Source: Patchwork Barents)

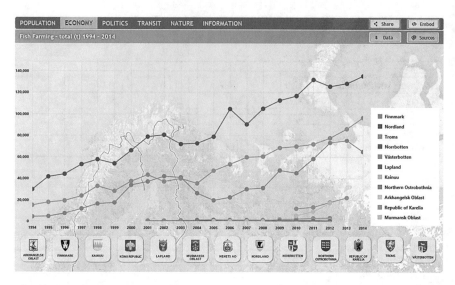

Fig. 4.2 The number of fish farms in the Barents area of Finland, Sweden, Norway and Russia (1994–2014). (Source: Patchwork Barents)

Local foods, as indicated by villagers, for example, in the village of Kylä, consist of fish, potatoes, meat and berries—they form the basis for the general nutritional requirements. In the village of Kylä, their meat mainly consists of reindeer meat, which is used by every household whether they own reindeer or not, but it can also include meat from cows and moose. A substantial seasonal variability was also described in the diet, where meat was mainly consumed during autumn and winter, though dried meat was also eaten in summer, particularly by men during work-related migration (herding, hunting and rafting). In many Northern areas, fish is deemed the second most important local food for nutrition, and in some villages it can be the most important food, especially for women. Fish are caught with rods and nets in lakes and rivers both in the summer and winter. Wild berries such as the cloudberry (*Rubus chamaemorus*), bilberry (*Vaccinium myrtillus*), lingonberry (*Vaccinium vitis-idaea*), raspberry (*Rubus idaeus*) and cranberry (*Oxycoccus*), are widely collected and eaten, they are available fresh during good seasons or con-served all year. Potatoes, strawberries (*Fragaria vesca*) and blackcurrants (*Ribes nigrum*) as well as some onions and root crops are commonly grown in family gar-dens. Mushrooms and herbs are picked from the forest and incorporated into meals but on a much smaller scale. Animals such as moose (*Alces alces*), rabbit (*Oryctolagus cuniculus*), fowl (*Galloanserae*) and waterfowl (*Aix galericulata*) are lower in terms of availability. Dairy farming also exists to produce a wide range of dairy products, including milk, cheese, yoghurt, butter and other fermented dairy products.

The cultural aspects attached to these foods can be a source of income to the locals. Foods found in the region can have a multipurpose use; for example, reindeer bones can be used as utensils, and other products. At the same time, reindeer fur can be sold in the market for additional income. These foods by definition would be considered local foods, but due to the cultural aspects that people in the area have attached to them such foods are also considered traditional foods. To keep the taste and flavor of country or traditional foods, it is desirable to engage novel technology that ensures a better utilization of these foods through preservation and processing techniques, which can eventually support the local economy.

4.2 Value Addition to the Traditional Foods in the Region

A food security study conducted in the year 2001, made a number of suggestions concerning sustainability in regards to food security. These suggestions were explored using legal tools, policies and frameworks coupled with models that would help place the focus on sustainable food security. One of the suggestions is to establish a structure for marketing local food as an avenue to reduce food imports and a community's dependence on food products manufactured in the South. This idea would contribute to generating a certain amount of wealth within the communities, not only for hunters or fishers who find a place to sell their harvest but also for the population as a whole. The local economy in the region is expected to also grow if small and medium enterprises engage in value addition to traditional foods. Currently, some northern communities are coming up with new initiatives that support their own branding and marketing of traditional foods such as Arctic ice cream, berries, meats, beer and similar products. These suggestions are important, but far more solutions that are concrete are needed for making the Barents region more food secure. The food security report edited by Gerard Duhaime did not provide definitive solutions but rather posed questions that could be useful in providing further sustainable food security to the region. The questions are as follows: "with respect to legal frameworks, which structural limitations could be the subject of change to foster the development and promotion of local food? With respect to international policies, how it is possible to generate an international acceptance to the promotion of activities related to the production of local food given a certain resistance to the commercialization of these products, or at the very least the major force of inertia that must be overcome to start up such projects in the Arctic-Barents?" (Duhaime and Bernard 2001). These are valuable questions that were posed more than 16 years ago and are still relevant today. The link between food and tourism has also been shown to be an economic driver in the region. Most tourists who visit the region are interested in the local and traditional foods (Havas et al. 2015). For example, Rovaniemi, the capital city of Lapland and the official hometown of Santa Claus is exploring economic opportunities to explore the fresh products of northern forests and waters. The city is the second most visited city in Finland after Helsinki, and most of the visitors are from abroad. In Lapland, sautéed reindeer meat served with mashed potatoes and fresh lingonberries or lingonberry jam is the most classic dish. In promoting the use of local ingredients in local foods, each spring, Rovaniemi hosts the 'Reindeer Chef of the Year' (*Vuoden Poro kokki*) competition, which draws renowned chefs from around Finland (Visitrovaniemi 2017).

Fig. 4.3 (**a**) Wild berries. Photo credit: Päivi Soppela. (**b**) Innovative value addition to wild berries. Photo credit: Biokia

Fig. 4.4 (**a**) Garden angelica, *Angelica archangelica*. Photo credit: Jouko Lehmuskallio. (**b**) Gel shot from angelica and nettle herbs. Photo credit: Arctic Warriors

The issues of promoting local food and even traditional food are important to set a meaningful standard for food security in the Barents region. Some good practices (from the Finnish Barents region) that have been successful in regards to value addition to berries, the angelica plant and reindeer meat are shown in Figs. 4.3, 4.4, 4.5, and 4.6 below.

Gamebirds such as capercecillie, willow grouse, black grouse and waterfowl are hunted for certain dishes as are elk and bear (Visitrovaniemi 2017). Berry picking in the forests and bogs is popular among the locals during the period from late July to September. Cloudberry is the crown jewel of Lapland's berries. Bilberries (northern blueberries), lingonberries and cranberries are also desired for consumption (Arctic Flavours 2017). They are made into juices, jams, sauces and liqueurs. Lapland's own potato, which is oval-shaped, like an almond, is also a delicacy. Other root vegetables include carrots and turnips (Visitrovaniemi 2017). Lapland angelica herb is used as flavoring for pies and ice cream. A dessert bread cheese (*leipäjuusto*) is a

Fig. 4.5 (a) Cloudberries (*Rubus chamaemorus*). Photo credit: Päivi Soppela. (b) Cloudberry jam. Photocredit: Meritalo

Fig. 4.6 (a) Reindeer foraging in the wild. Photo credit: Päivi Soppela. (b) Innovative value addition to reindeer as chips. Photo credit: Mainostoimisto Puisto Oy

local specialty that can be served with fresh cloudberries or cloudberry jam. Soft barley flat bread (*rieska*) is usually enjoyed with butter (Visitrovaniemi 2017).

For example, Rovaniemi city has many dining outlets; Marmaris pizzeria and a bistro (ROKA), shown in Fig. 4.7, are a common sight. In the background is the city's soccer stadium with its floodlight. Many tourists that visit the city enjoy sampling the local and traditional foods.

The foods that are available in the Arctic Barents region is further influenced by global trends, there is an interest in street vendors which appeals especially to the younger generations, and these new food sources incorporate social media into their promotional strategies as shown in Fig. 4.8.

Fig. 4.7 Marmaris pizzeria and ROKA bistro in downtown Rovaniemi. Photo credit: Dele Raheem

Fig. 4.8 Ravintola ROKA is expanding their food business operation this summer (2017) to the streets of Rovaniemi. Photo credit: Dele Raheem

Market days provide an important opportunity to become familiar with traditionally produced and local foods in different parts of the region. Popular market days include the *Rovaniemen wanhat markkinat* (Finland), *Jokkmokk marknad* (Sweden), *Markomeannu* festival (Norway) and the Pomor food festival (North West Russia).

Fig. 4.9 A tourist buying some local food products from a market festival during the St. Mary's Day festival in Hetta, Finnish Lapland. Photo credit: Dele Raheem

Fig. 4.10 Locally produced rapeseed oil at the 2016 Swedish Lapland (Pajala) market day. Photo credit: Dele Raheem

The St. Mary's Day festival at Hetta, Enontekiö, features a range of local and traditional foods and includes handcrafts made by the indigenous people.

One important way by which traditional foods are promoted in the region is through the opportunity to link food with culture. The artisanal value addition to these food products can be important ways to earn income for the local producers. Figures 4.9, 4.10, and 4.11 are a few examples of locally produced foods market.

Fig. 4.11 Locally produced jams and juice from berries at the 2016 Pajala market day. Photo credit: Dele Raheem

In order to innovate and bring benefit to local foods of high quality that end up in the market will demand a consistent supply of these foods. This will be best guaranteed with support to growers, processors and other stakeholders in the region.

Chapter 5
Food Security and the Arctic-Barents Communities

5.1 Arctic-Barents Communities

Many communities and areas within the Barents region are facing their own challenges concerning food security. The extent of these challenges varies, however, among the different areas and communities. The variations between communities and regions affect access to foods and their availability. Access to food and its availability also vary from year to year depending on the weather as well as on the consequences of climate change. For example, communities that are located along the Norwegian coastline are likely to face different threats to food security when compared to those situated inland, as in Finland or Sweden. Depending on the location and road infrastructure, some remote towns and villages may not have access to a nearby grocery store, leaving them with limited access to food and markets. Moreover, the economy and productivity of communities and regions differ as well, which affects access to store-bought foods.

Figure 5.1 shows the location of indigenous peoples in the circumpolar regions of the world. The indigenous peoples in the Arctic-Barents are on the lower part of the map while the upper part are of the Arctic regions of Canada and America.

The communities in Northern Finland might tend to focus on herding reindeer, whereas people living on the northwestern Russian coast may tend to fish. It is difficult to determine the full extent of the food security situation for every village and community in the Barents region due to the limited amount of research conducted on such circumstances. However, a number of scenario-building workshops were held in specific towns and villages, which gave us a general outlook in regard to certain communities in the region.

A series of four workshops across the Barents region with the aim to build visions of different local futures in the region under different climatic and socio-economic contexts were conducted using the same methodology and research questions. These workshops connected local change to global scenarios (van Oort et al. 2015). These scenario-building workshops were held in June and August 2015 in the

© Springer International Publishing AG, part of Springer Nature 2018
K. Hossain et al., *Food Security Governance in the Arctic-Barents Region*,
https://doi.org/10.1007/978-3-319-75756-8_5

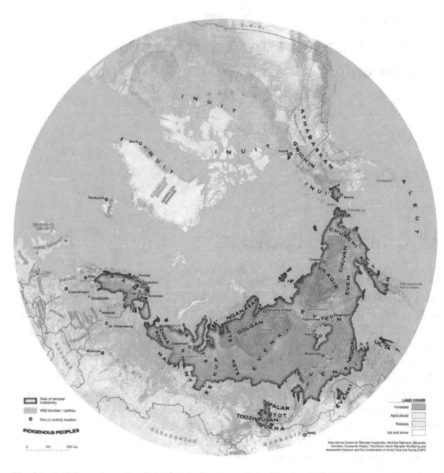

Fig. 5.1 Indigenous peoples of the Arctic-Barents region. (Source: ICR 2017)

communities of Kylä, Kirovsk, Inari, Bodø and Pajala to offer further insights into the present and future challenges that the communities and surrounding regions are facing (van Oort et al. 2015). While these issues are not specifically directed at food security, they do concentrate on a wide range of concerns that involve food security. These workshops aimed at developing a bottom-up participatory scenario-building process that would be relevant for actors in the Barents region. Such scenario approaches are said to be frequently used in decision-making when future developments are highly uncertain but the resulting decisions may have long-term implications (Raskin et al. 2002). During the workshops, participants addressed the issues raised through voting, where they decided on which drivers of future change have the biggest impacts and are the most uncertain and ranked them in order of importance. Many of the choices were closely related and therefore grouped into clusters. As a result, the way these workshops were constructed allowed for participants such as locals, professionals and researchers to gather and discuss longer-term questions surrounding the present and future challenges of their communities.

Pajala is a small community in northern Sweden of about 6300 residents, which had an iron ore mine open in 2012 but that went bankrupt in 2014 (Jarratt 2014). Despite the difficult situation, the community is still underpinned by tourism, electronic manufacturing, the service industry and forestry. The outcome of the workshop was produced in a final report, which presented several challenges that were discussed in this meeting. The report highlighted the Kaunisvaara mine in Pajala municipality as a future challenge for the Saami village. Before this mining project, villagers' activities, especially those related to the large grazing areas needed for reindeer herding, had not been affected by extractive industries (Nilsson 2015). In addition, the research showed that Pajala struggles with reducing food waste and finding ways of sorting rubbish for the elderly living in remote locations, which has also become an environmental issue. In ranking the drivers for future change, climate impacts were ranked 25th in importance and second in terms of uncertainty (Raskin et al. 2002). On the other hand, climate change, was ranked first in importance and fifth in uncertainty (Nilsson et al. 2015). This is unique considering the participants' ranking of climate impacts as much lower in comparison to climate change in general. The report suggested that one possible interpretation of this ranking by the participants could be based on a judgment that makes it relatively easier to adapt to climate change, although the participants ranked the uncertainty regarding impacts very high (Nilsson and Evengård 2015). The concluding remarks of the report shed light on Pajala in an international context where global demographics combined with climate change and other conflicts directly influence Pajala and similar communities in northern Fennoscandia. It was noted in the report that increasing numbers of people are fleeing from these conflict areas and the hostile climate where they lack basic human securities such as food, water and shelter (Nilsson and Evengård 2015).

Similar scenario-building workshops were conducted to compare two communities—Kirovsk and Bodø in the Barents region of Russia and Norway, respectively. The workshop addressed the following questions: What does the future look like from the perspective of municipalities in various locations in the Barents Region? What climatic, social and environmental challenges might there be, and how might local people respond (van Oort et al. 2015)?

The city of Kirovsk is located in the central part of the Murmansk region, and its main economic activities involve the extraction and processing of apatite ore and tourism. The Murmansk Oblast is an area that covers about 150,000 km^2 with a population size close to 800,000, of which approximately 28,625 live in Kirovsk. The majority of the population lives in urban areas, whereas <10% live in rural areas. Murmansk Oblast is very rich in natural resources and has deposits of over 700 minerals (van Oort et al. 2015). The largest industries are metallurgy (36.6%), electric power production (22.9%) and the food industry, including fishing (13.7%) (van Oort et al. 2015). The oblast has a 41% share of the total Russian marine transport market (van Oort et al. 2015). Similar to the Pajala workshop, each participant in the Kirovsk workshop was given four "post-it" sticky notes on which to write suggestions regarding what the important driving forces for future changes are that can have economic, environmental and social consequences for the Murmansk region over the next 30–50 years (van Oort et al. 2015). In ranking these drivers, environmental conditions were ranked second in importance and fifth in terms of uncertainty, while climate change was deemed sixth in importance and first in terms of uncertainty. However, it was difficult to compare the results obtained from the Bodø and Kirovsk communities, as each workshop had a different group of clusters and challenges that they believe their community will face in future decades.

In Bodø, Nordland, of Norway, the situation was quite different. Nordland covers an area of about 38,500 km² and has a population of around 240,000. The population has been decreasing by about 1.6% yearly over the last 10 years (van Oort et al. 2015). The Bodø municipality has 50,000 people and is historically known for its fishing roots, which can be traced back many centuries (Visitbodo 2017). Today, Bodø is an important trade center for fish, as fisheries and aquaculture remain key industries, but there is also petroleum exploration and tourism. The Bodø workshop had 23 participants representing local, regional and sector-specific perspectives and included both practitioners and researchers. The main discussions involved many topics; however, in ranking the challenges through different clusters there were some surprises. "Climate change and impacts" were voted by the participants as second in importance and fourth in uncertainty (van Oort et al. 2015). One cluster that had not been mentioned in the other workshops was "Food security," which was ranked eighth in importance and tenth in uncertainty (van Oort et al. 2015). Possible explanations could be that the participants were concerned about climate change impacts and its effects on the fishing industry, which could lead to further impacts on the community and its food security situation. The report mentioned that, from a global perspective, the people of Nordland are lucky because they have adequate food and do not have too many climate-related problems (van Oort et al. 2015). However, we find this interesting, as no other community in these scenario-building workshops had brought forth the food security concern. The report also mentioned that the demand for food, water and energy will grow, fueled by changing economic conditions in the developing world, which may impact the food security in such communities in the Barents region (van Oort et al. 2015).

There are additional data that show a price hike in the overall price for a basket of food in the Russian areas of the Barents region (Staalesen 2015). The graphical data show the average cost of a basket of food for five regions in northwest Russia: the Republic of Karelia, Komi Republic, Arkhanglesk Oblast, Murmansk Oblast and Nenets Autonomous Okrug (AO). According to the graph, there has been a substantial price hike of as much as 2% in the Nenets AO region from November to December 2014. According to Staalesen (2016), Murmansk, the biggest city in the Arctic-Barents, witnessed monthly prices on basic foodstuffs such as frozen fish, eggs and beef increase more than 8%, 12% and 6%, respectively. If these prices are increasing substantially in the larger communities, one can only imagine the changes and price impacts on smaller or rural places. This also shows how the impact of the world market affect northern communities, as Russia was hit by a number of sanctions in 2014 that have impacted trade and caused inflation within the country (Staalesen 2016).

The scenario-building workshops, nevertheless, showed that communities in the Barents region are not homogenous; although some of them face similar kinds of challenges, they still remain largely different. However, the variations in terms of challenges are mostly due to the impacts resulting from climate change, hence this being a common theme among these communities. Climate change remains a threat to food security among the communities in the region. It is expected that the effect of climate change will, in the future, also impact food systems, resulting in challenges to food security.

5.2   Food and Community Connections

Food is a special way of connecting people, families, groups, communities and regions. Food remains a common bond among all human beings on this planet, and it is needed for survival (Welsch and Vivanco 2015). Indigenous systems were historically associated to people; place; culture; traditional and modern markets; food systems; physical, social and mental health; colonial histories and environmental and climate change as well as governance (ICR 2017). In addition to the physical and health aspects of food in the Barents region, as stated earlier, food entails cultural, traditional and spiritual dimensions. Throughout all the dimensions that are associated with food, the characteristics of the community are visibly emphasized. In the Barents region, food has a unique way of bringing together people and communities. The unity around food practices, in particular in relation to indigenous peoples, takes place throughout the whole process of hunting, gathering, preparing, trading, herding and selling (ICR 2017). All of these practices are associated with the traditions and culture that have prevailed for generations in the region, not only for indigenous peoples but also for non-indigenous populations (Nuttall et al. 2005).

Generally, the unique cultural practices built around food practices provide a sense of community in the region. Although there is no set definition in regard to who forms a community, the phenomenon can be explained according to the work of Mason (2000) as a group of people who share a range of values and way of life, who identify with the group and its participants and who recognize each other as members of that group (Mason 2000). Food is a common bond among communities, especially among the traditional Arctic-Barents communities, where gathering, hunting and eating food take place mostly in family and community atmospheres. This community connection to food strengthens food sovereignty through incentives to provide more ownership to the local community in terms of the creation of a sustainable food system. A sustainable food system is dependent upon the practice of the ecologically sound production of food, to which community voices are integral in regard to expressing their choices for healthy and culturally appropriate foods (Nyeleni 2016).

Food security, in its narrowest sense, does not distinguish between where food comes from and the conditions under which it is produced or distributed. The targets set by national food security policies are often met by sourcing food produced under

Fig. 5.2 Food culture and traditional knowledge of Arctic indigenous people. (Source: EALLU Yakutsk. Photo: Victor Zabolotskiy [ICR 2017])

environmentally destructive and exploitative conditions and are supported by subsidies and policies that destroy local food producers but benefit, for example, agribusiness corporations (Barkin 2016). As food sovereignty, which we have referred to elsewhere in this book, is the indicator for the promotion of food security in the Arctic-Barents region, we stress that community connection has to be established by emphasizing ecologically appropriate production, distribution and consumption. At the same time, we see that social-economic justice in local food systems as a way to tackle hunger and poverty, and to guarantee a sustainable supply of food for all peoples, is contributory to food security. Such a system advocates trade and investment that serve the collective aspirations of society. It promotes community control of productive resources, agrarian reform and tenure security for small-scale producers. Furthermore, the system as such also promotes agro-ecology; biodiversity; local knowledge; the rights of peasants, women, indigenous peoples and workers; social protection and climate justice.

Figure 5.2 is a picture from the EALLU project workshop on the importance of traditional knowledge and food culture among the indigenous peoples of the Arctic. EALLU is a Saami word, which translates to 'herd' in English. The Arctic Council Sustainable Development Working Group endorsed the EALLU project at the end of the Canadian chairmanship in 2015. The goal of the EALLU project is to investigate reindeer herders' food culture through the lens of traditional knowledge, adaptation to climate change and youth. The EALLU project will combine academic work, education, seminars, food culture across Eurasia (ICRH 2017).

Community connection to food is explained by the livelihood practices conducted by each of the specific communities in the region. The reindeer herding practice is a common activity throughout the whole region and is well-documented as a community-based activity, usually undertaken in "*siidas*" (Mustonen and Jones 2015). As discussed earlier, the herders often work in partnership within the community, where the members have individual rights to resources but help each other with the management of the herds. According to Bjørklund, the *siida* represents a flexible cooperative unit between people and animals (Bjørklund 2004). This type of group connection is important for food security in the regional context because of the historical practice of food sharing among family members within the community. This assures the principle that everyone has enough food to eat and that no one will starve.

The advantages of such a group atmosphere that involves sharing are many, but it is also necessary to share food within the society, because stocks of fish, game animals, valuable fur from animals and other resources are unevenly distributed within the districts (Dana and Åge Riseth 2011). A similar trend occurs with land ownership as well; instead of dividing land, it was more beneficial for the children and future generations to simply pass down the lands for sharing and maintaining their way of life (Dana and Åge Riseth 2011).

In regard to community connection, the essence of reindeer herding is such that an individual can own reindeer, but one cannot herd reindeer alone. Therefore, the Saami view herding as an expression of traditional cultural values, and they value the communal activity revolving around the family. Many of the activities in connection to food in the Barents region are completed by utilizing and assembling community members around these practices. Although reindeer herding in this regard is used as a primary example, hunting is another important one. In some areas of the region, hunting used to be traditionally done in groups. According to Juha Joona—a researcher at the Arctic Centre—hunting was traditionally done as a collective exercise, where all males in the village would have to help out or participate in some way. For example, after the deer had been hunted and killed, it would then be divided among those who had participated in the hunt. He also mentioned that hunting at this time was a social gathering and based on the social aspect of community (Joona 2016a, b: Interview). Activities such as herding, hunting, fishing and gathering, especially among indigenous societies, were vital for maintaining social relationships and cultural identity. They define a sense of family community and reinforce and celebrate the relationships between indigenous peoples and the animals and environment upon which they depend (Callaway 1995; Nuttall 1992).

Apart from gathering and producing food connecting people, the trading and selling of such foods also connect people. As previously discussed, this region has been historically known for its trade among communities and large areas over many centuries. Trade was mainly done in markets spread throughout the community; this would allow people to attend these markets at different times throughout the year (Müller and Petterson 2001). In fact, some of these markets still exist today, such as the Jokkmokk winter market (Müller and Petterson 2001). The trade of goods involved fur, equipment and food. It has been said that there was active trade and social interaction among Russians, Saami, Karelians, Vepsians, Norwegians and Finns from 1809 to 1905. Although the influence of globalization has changed this in many ways, the trade of food products is still heavily ongoing in the Barents region (Elenius et al. 2015). Before the recent sanctions were put on Russia in 2014, the Russians traded a number of goods with Scandinavian countries (Finland is the only EU country that shares a border with Russia), among which foods such as dairy products, fish, and other foods sourced from animals were most common (Elenius et al. 2015; ETLA 2016).

The community connection can also be demonstrated through other activities, such as tourism. Tourism has been a growing industry in the Barents region in recent years, especially in the Scandinavian countries (Kohllechner-Autto 2011). Tourists come from all across the world to experience the vast Arctic-Barents landscape and to familiarize themselves with its unique cultures and cultural diversity. The industry has developed experiences that further connect communities and locals with new and exciting opportunities for tourists. In recent years, there has been more marketing of traditional foods and local foods in the area, especially in Lapland (the northernmost region of Norway, Sweden and Finland), where tourists come to experience reindeer meat, fish, wild berries and other local foods (Kohllechner-Autto 2011). This has created opportunities for local people to sell berries in farmers' markets as well as internationally. Additionally, there are a wide range of tours involving hunting, fishing and berry picking.

For example, a woman in Nellim village located next to the Russian border of the Inari municipality runs her own food service enterprise. She is a professional cook and has written several books about the Skolt Saami food tradition specializing in foods that are either produced or collected from the village area. She offers services for tourists by integrating the food culture practiced in her locality. Professor Satu Miettinen has carried out field studies in which she interviewed the woman. According to Miettinen, after the woman has given a tour of the village and after showing the tourists all the sites, she arranges for them to have dinner at her home. At dinner time, she explains the history of traditional Saami Skolt dishes to her guests and the origin of the different fish, vegetables, meat, berries and herbs used in the dishes (Miettinen 2006). This creates a special experience for tourists to see how local villagers live their daily lives.

The Nellim Orthodox Church dedicated to the Holy Trinity and to the memory of Trifon Petsamolainen organize annual pilgrimages. Before the pilgrimage, local villagers provide dinner for tourists visiting the area, where they serve traditional foods. The dinner is often accompanied by role-plays that are related to the history and traditions of the village. These are just a few examples of the events, tours and festivals in the region promoting the community aspect of the practice of food culture.

All across the Arctic-Barents it can be observed that many indigenous communities are increasingly being characterized by multi-plural activities, so that money can be earned either through full-time or part-time paid work, seasonal labor, craft-making, commercial fishing or other pursuits, such as tourism that supports and supplements renewable resource harvesting activities (ICR 2017). The community accomplishment of such tasks in relation to food practice offers societal values through which people in the region find their way of living, spiritual satisfaction and cultural affinity. According to AMAP, the sharing culture and exchange of country or traditional food within a community is an important element of social wellbeing and is intrinsic to local culture (AMAP 2009). These activities and the knowledge about these activities are important for the region's population and for their future generations for them to maintain traditionally developed social and cultural sustenance. By maintaining the relationships around food practices, a community also finds its way to establish, or to help build, a communal identity, the protection of which elaborates other components of human security, such as community security.

5.3   Food and Indigenous Peoples

Indigenous peoples, as frequently referred to in this book, are spread out across the whole Barents region. They have a history of inhabiting the region for centuries. For these peoples, food is a unique and important aspect of their everyday lives. The relationship between food and the indigenous peoples in the region has been well documented in history. They have engaged in many forms of traditional economies throughout the centuries. They remained dependent on the natural resources and foods available in their natural surroundings (ICR 2017). In fact, their connection with food extends beyond physical survival incorporating nutritious value for health to overall wellbeing in connection to the maintenance of culture, spirituality and identity. Before grocery stores and supermarkets, indigenous peoples used their natural environment for the collection of foods. They used traditional knowledge for the preservation and sustainable maintenance of the food system (Magni 2016). They are regarded as an excellent example of food secure communities, despite not having the concept or an understanding of the definition concerning food security at that time. This was accomplished using traditional methods for hunting, gathering and fishing. The mastering of these skills and processes took place over the course of many generations. Noreen Willows describes traditional food as those culturally accepted foods available from local natural resources that constitute the food systems of indigenous peoples (Willows 2005).

As discussed above, traditional foods remain important for indigenous peoples for a healthy and sustainable lifestyle. The methods of collecting foods provides indigenous peoples with the extensive physical exercise needed to maintain a well-balanced lifestyle. Beyond the health aspect of these activities, Woodley et al. (2006) believes that traditional foods and food practices are deeply intertwined with culture and value systems and play an important role in religious ceremonies and spirituality as well as in songs, dances and myths. Throughout the entire food system, many stories are told, shared and passed on through families to new generations. Stories describe in detail the connections between land, the environment, food, community, family, religion and animals. Woodley demonstrates that ceremonies; oral traditions, such as stories, songs and oral histories; and other cultural practices, such as reciprocity, are important cultural elements in the maintenance and transmission of the knowledge and practices of traditional food and agro-ecosystems (Woodley et al. 2006). This transmission of knowledge is important for the continuation of traditional activities, such as teaching the next generation hunting, herding and gathering practices. Woodley et al. (2006) explained that the loss of cultural practices creates a disconnect in the relationship between culture and traditional food systems. Therefore, activities through which people identify themselves with their culture and natural environment as well as the knowledge and use of traditional food systems to improve health build community support and engagement for holistic health and wellbeing (Kuhnlein and Burlingame 2013). It is believed that the longstanding dependence of contemporary indigenous societies on hunting, herding, fishing and gathering continues for several critically important reasons (Nuttall et al. 2005). Nuttall et al. (2005) highlighted two main reasons for the importance of accessing customary and local foods for economic and dietary reasons. Firstly, many of these foods are deemed nutritionally superior to the foodstuffs that are currently imported. Second are the cultural and social elements associated with hunting, herding and gathering. In addition, the communal processing, distribution and consumption of foods as well as the celebration around the practices are also important to them. These food activities are connected in different ways among all indigenous peoples in the region. Across most of the Barents area, a general trend is observed in relation to a political move toward self-determination, which explains the need for greater voices in matters that critically affect them. Such political move relates to the food system as well the accommodation of the voices of the indigenous peoples and their participation in the promotion of traditional food production, since the system highlights community preferences.

5.3.1 Saami

The Saami homeland stretches from central Norway and Sweden through the northernmost part of Finland and into the Kola Peninsula. There are approximately 90,000 Saami, of which the Norwegian Saami constitute the largest group, numbering approximately 50–65,000 people, followed by Sweden with 20,000, Finland with 8000 and Russia with 2000 (Barents Info 2016a, b, c, d). The Saami have their own history, language, culture, livelihood practices and ways of life on which their identity is built. They participate in a number of traditional economies, such as reindeer herding, fishing, hunting and gathering. However, many Saami also work in the cities and have other day jobs as well (Revolvy 2017). Reindeer herding is exclusive for the Saami in Norway and Sweden, but in Finland non-Saami people can also herd, and the same applies to Russia, where both Saami and non-Saami are allowed to herd reindeer (Forrest 1997). It has been noted that Saami who still engage in reindeer herding, as a surrogate determinant of a traditional lifestyle, have a different risk for certain diseases such as musculoskeletal and cardiovascular diseases, compared to those not engaged in reindeer herding (Hassler et al. 2008). Alastair Ross also pointed out that the traditional Saami diet consists of a high amount of animal products, particularly from reindeer, and a low amount of grains, fruits and vegetables (Håglin 1991). Some believe they have adapted to this diet, high in protein and fat and low in carbohydrates, as part of their traditional lifestyle of hunting and reindeer herding (Håglin 1991; Lehtinen et al. 1998). In addition, it has also been noticed that other dietary characteristics of the historical Saami and present-day reindeer-herding Saami were/are higher intakes of fat, blood and boiled coffee and lower intakes of bread, fiber and cultivated vegetables compared with present-day non-Saami (Nilsson 2012). Even though not all Saami participate in reindeer herding, they do participate in other traditional activities such as fishing, berry picking and hunting (Ross 2009). However, much like other indigenous peoples in the region, the increased colonization of Saami areas coupled with increased contact with other Nordic peoples make the region more "Westernized," leading to the more Westernized lifestyle of the local populations. The opportunity to be self-reliant on traditional foods requires being able to utilize the available aquatic resources within the Saami region. For example, the 361-km river, the Teno, which runs through the Saami area of Finland and Norway, supports the largest Atlantic salmon stock in the world, and it is the most prolific salmon-producing river in both Finland and Norway. The recently adopted Teno Fishery Agreement, which took effect in May of 2017, has been widely disputed among the Saami from the viewpoint of their human security to food and food sovereignty, given that the Saami parliaments in Finland and Norway were not included in the process of negotiation (Yle Saami 2017).

5.3.2 Komi

The Komi, a Permic-speaking people, live mainly between the Pechora and Vychegda rivers, southeast of the White Sea, in the northern European area of Russia. They are of the Finno-Ugric branch of the Uralic family. The Komi are known to comprise three major groups: the Komi-Zyryan of the Komi Republic; the Komi-Permyaks (or Permyaks) of the Komi-Permyak Autonomous Okrug (district) to the south; and the Komi-Yazua to the east of the okrug and south of the Komi Republic (Barents Info 2016a, b, c, d).

According to the Russian Federation's legislation, groups including over 50,000 peoples are not recognized as indigenous. Therefore, the Komi, with a population of 901,189 and an average of 2.17 people/km^2 (RFSSS 2011), are not officially recognized as indigenous peoples, but they have the status of indigenous people in the legislation of the Komi Republic (Barents Info 2016a, b, c, d).

The Komi people do not live in the Republic of Komi only. They also live in the Nenets Autonomous Okrug, the Arkhangelsk Oblast and the Murmansk Oblast. Some Komi groups are also found in the Siberian part of Russia. The Komi are also a minority in all these territories, even in the Komi Republic, where they make up only 23.3% of the whole population (Habeck 2002). As in many other regions of Russia, the distribution of ethnic groups in the northeast of European Russia does not coincide with the administrative boundaries of their ethnic home-lands (O'Loughlin et al. 2007). For example, the eastern part of the Nenets Autonomous Okrug is inhabited by both Komi and Nenets in almost equal num-bers. Russian settlers are known to have lived along the Pechora for more than 500 years (Habeck 2002).

The Komi have been nominally Russian Orthodox since the fourteenth century (Britannica 1998). The severity of the harsh climate and the inaccessible geographic location kept them culturally isolated until after World War II (Nuttall 2012). The economic activities of the Komi vary from reindeer herding, hunting, fishing and lumbering in the north (with a mining center above the Arctic Circle at Vorkuta) to agriculture, industry and mining in the south (Britannica 1998). The agriculture in the area is very small. Although 24% of the population lives in rural areas, only 2% of the population in the region is engaged in agriculture. In the late summer and fall they also pick mushrooms and berries (Interview with Bruce Forbes 2016). Many of the Komi people also have gardens—"*dacha*"—to supplement the demand for food, which actually remain from the Soviet times, where they grow herbs and other veg-etables (Caldwell 2011).

5.3.3   Nenets

The Nenets—an indigenous group—mainly live in rural areas and villages spanning the Nenets Autonomous Okrug, Arkhangelsk Oblast, Komi Republic and Murmansk Oblast in Russia (Barents Info 2016a, b, c, d). The Nenets are the most populous indig-enous peoples in Russia, with a population of 44,000 (according to a 2010 census) (Rohr 2014). Some consider the Nenets to be among the world's last true nomads (Eshelby 2015). Their diet is similar to that of other indigenous peoples in the area, with reindeer meat being the most important. It is eaten raw, frozen or boiled, together with the blood of a freshly slaughtered reindeer, which is regarded rich in vitamins (Eshelby 2015). Eating raw meat occupies a special place in the Nenets traditional diet, because raw reindeer meat and blood contains a lot of the vitamins and minerals needed to survive and be healthy in the harsh conditions of the North (ICR 2017). The Nenets also eat plenty of fish, such as white salmon and muksun, a silvery-colored whitefish. They gather mountain cranberries during the summer months. The Nenets have tradi-tionally relied on proteins and fats as their main sources for nutrition (Yakovleva 2005). However, studies carried out by Natalia Petrenya in Nelmin-Nos and the city of Arkhangelsk found the current levels of fish consumption among the residents of Nelmin-Nos to be inadequate. The average fish consumption was equal to approxi-mately 1.25 portions of 150 g/portion in a week. The residents of the city of Arkhangelsk consumed approximately 2.25 portions of a 150 g/portion per week on average. Changes in the nutrition of the indigenous population may lead to an increase in obesity and, later, to chronic diseases (Murphy et al. 1995). Their main traditional economy is based on the selling of reindeer meat. Other than just selling reindeer meat for commercial purposes, it is also used as a source of food, shelter, clothing, transport, spiritual fulfilment and means of socializing (BBC 2014). Reindeer hide is used for making coats, lassoes are crafted from reindeer tendons and various tools and sledge parts come from their bones (Survival International 2016). The Nenets have rich knowledge about the extended preservation of reindeer meat in traditional ways; they depend on various factors such as time of year and weather conditions. For instance, in autumn, Nenets often slaughter several reindeer a day. This is because in this period reindeer skins are at their best for use as clothing (ICR 2017). From skins harvested in autumn, Nenets sew malitsa (for men), pany (for women) and sovok (for men—with fur on the outside, which is used as an additional top cover in the cold winter weather). All of these traditional garments are still used in everyday nomadic life and indeed are central to the maintenance of the traditional herding way of life in Yamal (ICR 2017).

Traditional hunting was carried out in the region with traps, and snares were used for polar foxes; but nowadays, with no market for the polar fox, hunting is done more for sport and on a limited scale to add a little variety to the meat diet (BBC 2014). For Nenets with smaller herds, fishing is of particular importance. It accounts for most of their income, especially during the summer months when meat cannot be stored. During the winter months, the Nenets fish through ice holes by using a large net that is set underneath the ice.

5.3.4   Veps

The Veps live in small villages and remote parts of the Republic of Karelia and in the Leningrad and Volodga oblasts. They are recognized as indigenous people according to a statute issued in 2000. In a 2002 census their number was 8240. However, this number seemingly decreased by 2010—the year when they were counted as numbering 5936. They are a Finnic people who formerly occupied a large area in the south and east of Karelia (Taagepera 2011). According to the "Red Book of the Peoples of the Russian Empire," Viires (1993), the present-day habitat of the Veps, lies between the lakes of Ladoga, Äänisjärv and Valgjärv, where they live in three separate groups. The first group—the Äänis—or Northern group is situated in Karelia, near Äänisjärv, to the south of Petroskoi. The Äänis-Veps call themselves lüdinik or lüdilainen. The second group—the Central Veps—is the most numerous group; they live in the St. Petersburg region of the Russian Federation on the river bank in Oyat. The third group is the Southern Veps, who live in the eastern part of the St. Petersburg region on the northwestern edge of the Vologda province on the River Leedjõgi. The Southern and Central Veps have infrequent contact with each other. The Northern Veps are separated from the other groups by the River Süväri and the interpolation of Russian settlements (Viires 1993).

The Veps mostly live in small villages in remote parts of all these regions, and only 37.5% of Veps people consider Veps as their native language. The Veps are engaged in agriculture, stone carving, forestry, pottery and other handicrafts (Taagepera 2011). They were historically described to be engaged in slash-and-burn agriculture. Hunting and fishing were also essential for the Veps; these are mostly conducted in nearby water, which is abundant in fish, and forests, which are full of game (Barents Info 2016a, b, c, d).

In this area of Russia, reindeer herding is not particularly practiced. Mainly the practices of agriculture, hunting and fishing serve as means for traditional economies. The Veps also hold other jobs in society for income instead of relying solely on these traditional economies. In 1703, Peter the Great founded a metalworking and munition factory near Äänisjärv. From then on, work in the factory became a source of income for many Veps of the Aunus province. Formerly, many Veps had migrated for work outside of the region. Many of them travelled to other parts in Russia as well as to Finland and Estonia (Viires 1993).

5.3.5 Pomors

The Pomors hold an overwhelming majority in the Arkhangelsk Oblast, where 6500 people defined their nationality as Pomors in 2002 (Barents Info 2016a, b, c, d). They live along the shores of the White Sea; their main occupation historically was sea fishing for herring, navaga and salmon (Elenius et al. 2015). They also became merchants, seafarers, explorers and naval seamen and officers, which was advantageous for initiating trade early on in the region (Barents Info 2016a, b, c, d). According to the library records of the University of Tromso, Norway, the so-called "Pomor trade" was of great importance for the economic and cultural development of northern Norway and of Arkhangelsk during the seventeenth and the eighteenth centuries (Revolvy 2017). The Pomor trade was based upon the bartering of fish products and grain. Flour and grain were brought to northern Norway by the Pomors. Rye flour—"Russian flour"—was the main commodity that was bartered for various kinds of fish, such as flounder, cod and salmon (Elenius et al. 2015). Because the Russians were also interested in fur products, their trade had great significance for the Saami as well. The trading between the Pomors and the northern Norwegian population was extensive, and this was particularly the case during the years of the gunboat wars in the early nineteenth century. During the nineteenth century, more than 300 Pomor ships visited the northern regions of Norway annually, when Russian trade reached its greatest extent (UIT 1999).

5.4 Food and Non-indigenous Peoples

In the Arctic-Barents, like most of the areas of the world, indigenous peoples form only a minority. Over the last few centuries, there has been a lot of in-migration to the Barents region from the southern parts of both Nordic countries and Russia

(Castberg et al. 1994). During this in-migration process, non-indigenous peoples in many cases have adapted to a lifestyle similar to (if not the same as) that of the indigenous population. Traditional culture, in particular concerning food practice, has influenced these other populations. During this time, they developed traditions, techniques and methods surrounding food and food activities. As a result, regional changes do not only affect the indigenous peoples, they also pose a threat to non-indigenous populations in many different ways. While many of them do not practice the traditional livelihoods, compared to what most indigenous peoples do, they do, however, rely on foods found from natural resources. As a result, any threat to food security becomes similarly detrimental to the non-indigenous population as it is for the indigenous peoples.

Throughout the years many of the traditions, which used to be regarded as indigenous culture, have been learned by the non-indigenous inhabiting the region. For example, the practice of traditional agriculture, such as farming, was taught to many of the non-indigenous locals. In many of the countries in the Barents, these locals possess rights that were generally regarded as emblematic of indigenous culture, such as reindeer herding. Because of the small amount of research in the Barents region in regard to practices held by non-indigenous population groups, it is hard to gauge the similarities and differences between these peoples in terms of food practices (Ross et al. 2006). Yet, as stated earlier, the non-indigenous population, in particular in Finland and Russia, take part in reindeer herding practices (ICR 2017).

In Finland—where the population includes indigenous Saami and ethnic Finns—reindeer herding is the lifestyle and occupation of about 1000 families, who rely on approximately 200,000 reindeer (Hukkinen et al. 2003). More recently, reindeer herding has been regarded as an old and impressively adapted livelihood supporting a unique cultural continuity of both the Saami and Finnish populations in northern Finland (Heikkinen 2006). Reindeer herding in Finland has developed rather differently compared to its Scandinavian counterparts; it is based on the "*paliskunta*" system rather than that of the Saami villages, and the majority of the herders are Finns (Heikkinen 2006). While there might not be a whole lot of difference between the traditional activities of both indigenous and non-indigenous populations, the reasoning and motivation behind participation in these traditional activities are different (Dana and Åge Riseth 2011). The reasoning was highlighted in a study conducted by Dana and Riseth, in which they compared the indigenous and non-indigenous herders in Finland. The participants who identified themselves as ethnic Finns viewed their self-employment as an individualistic form of entrepreneurship, and they focused their discussion on matters related to financial capital and profit (Dana and Åge Riseth 2011). In contrast, Saami respondents claimed that the causal variable behind their herding was the maintenance of a cultural tradition and not necessarily limited to the maximization of financial profits. The Saami people have always had collective ownership—that is, land belongs to the group in the form of the "*siida*," which involves collective social capital (Dana and Åge Riseth 2011). This ensures that stocks of fish, game, animals for fur and other resources are evenly distributed (Haetta 1996). Furthermore, reindeer herding skills that are acquired on the job—such as human capital, attitudes, beliefs, customs, habits, interests, lifestyle and traditions—are part of the cultural heritage passed from one generation to the next. Even though both the indigenous Saami and ethnic Finnish populations have lived in Finland for many centuries, Dana and Åge Riseth (2011) pointed out that it was clear that each ethnic group had, and may continue to have, their respective values and distinctiveness.

Non-indigenous peoples in this region do many of the activities revolving around food in ways similar to those of the indigenous groups. However, their motivation and reasoning for doing such activities may differ—perhaps not for all, but for most. Indigenous peoples have a cultural and spiritual connection to food, which is one aspect of food security, whereas all other aspects of food security are actively practiced by non-indigenous populations, too.

Chapter 6
Issues of Food (In) security in the Barents Region

6.1 Changes in the Barents Region and Their Impacts on Food Security

The Barents region has transformed a great deal over the past decades after the collapse of the Soviet Union, which put an end to the Cold War. Since then, the promotion of cooperation within the region has been imbued with a new enthusiasm among the actors to make the region nuclear free as well as a zone of peace. The focus has now shifted to the most pressing challenge that the region faces—climate change. Consequently, environmental sustainability as a response to the challenges of climate change and economic globalization was placed as a priority in the building of strengthened cooperation. However, these changes are rapid, and they have resulted in a volatility that brings new challenges, resulting in adverse effects on many dimensions of human security. Among these challenges, food insecurity is one of the most serious threats, since the regional transformations affect the quality and quantity of food.

After the Cold War, countries in the European Arctic-Barents region were determined to initiate cooperation on common issues related to sustainable development, and an enormous effort on environmental cooperation was emphasized in the Barents region. The Kirkenes Declaration was signed in 1993, leading to the establishment of the Barents Euro-Arctic Council (BEAC). The strategies of BEAC not only excelled in environmental cooperation, but the council also enhanced collaboration in the areas of economy, science and technology, regional infrastructure, indigenous peoples, human contacts, cultural relations and tourism (Pettersen 2002). It was suggested that the problems in the Barents region are cumulative in nature, as problems caused by different sources accumulate to form the core of the environmental cooperation in the region (Sellheim 2011). Due to enormous amounts of environmental degradation in the region from previous decades, Sellheim (2011) explained that private corporations have invested insufficient amounts of money into clean production and have thus far contributed to the large amounts of emissions stemming from the industrial complexes on the Russian side of the Barents region (Sellheim 2011). Environmental cooperation in the Barents Euro-Arctic Region (BEAR) was viewed positively by some, but with further economic cooperation leading to increased production and tourism coupled with insufficient amounts of money toward clean production in Russia, this has led to further implications of climate change and food insecurity in the region. The increased growth and investment has not occurred only in Russia but also in the whole Barents region within the last decade, especially in the areas of biofuels, mines and nuclear and aluminum plants. In 2010, it was estimated that 114 billion euros were set to be invested in the industrial development of the Barents region over the next decade (Nilsen 2010). These investments, when divided between individual countries, leave northern Finland with 25.6 billion euros and northern Sweden with 17.5 billion euros. Norway received the lowest with 9.5 billion euros, and finally Russia's Kola Peninsula came out on top with planned investments of 62 billion (Nilsen 2010). Growing investments in this region may lead to further environmental degradation, which will affect the climate with further impacts on food security in the area.

The frequently changing political situation and regulation of trade across borders also has an influence on the food security in the region. Nowadays, more people in the Arctic-Barents rely on market and imported foods rather than traditional foods, and this leaves them particularly vulnerable to market forces. The issue of food security has guided Russia's food policy since 2010, whereby the government combines intervention in the form of assistance for domestic production while simultaneously restricting market access (Wegren et al. 2016a, b). In 2014, the EU introduced sanctions on Russia for what the EU considered the illegal annexation of Crimea and Sevastopol by Russia (Szczepański 2015). This resulted in a number of sanctions on Russian officials and companies, leading to a decline of imports and exports of goods between the EU and Russia. Later in the same year, to counteract these measures, Russia introduced its own set of sanctions on the import of agricultural products, raw materials and food originating in the countries that had imposed sanctions against Russian companies and individuals (Szczepański 2015).

The Russian government recently extended its sanctions until the end of 2018 after the EU extended its economic sanctions against Russia (ABC News 2017). The Nordic countries as well as the Baltic States suffered by losing 10–22% of food exports due to this Russian food embargo (TASS 2016). In addition, Finnish food exports have dropped by 24.5% over the last 2 years, given that Finland used to export almost 20.9% of its food to Russia. Norwegian exports to foreign countries also dwindled—by 11.3%—including 10.1% due to the loss of the Russian market (TASS 2016). Foods that were banned include meats, sausages, fish, seafood, vegetables, fruit and dairy products (TASS 2016). These controls within the last few years have reduced the export of these banned foods by companies that market them on both the Barents and Russian sides, especially the export of products that are commonly found in the Barents region such as fish, dairy products and meat. These measures have seen increased costs of basic foods and even a shortage of foods in some cases. For example, on the Russian side, there was a price hike for a normal basket of food due to inflation. As described earlier, the price in the Nenets AO saw as much as a 2%

increase between November and December 2014 (Staalesen 2015). Due to food inflation, Russian consumers changed the structure of their diets and their buying habits
and began to purchase less expensive foodstuffs; as food purchases took up more of
the average household budget, many households started growing more of their own
food; and Russians began to prefer domestically produced food that was usually
cheaper than imports (Wegren et al. 2016a, b).

As a result of the 2014 food embargo, Russian food security policy has been
explicitly driven by import substitution (*importozameshcheniye*), which is intended
to replace imported food with domestically produced food (Wegren et al. 2016a, b).
There was an improvement in domestic food production in Russia through this policy of import substitution, and food is used as a foreign policy weapon. The list of
embargoed nations has grown since 2014, and the Russian government has empowered *Rossel'khoznadzor*, the agency in charge of food inspections to enforce its food
embargo (Wegren et al. 2016a, b).

The initial framework and structure of the Barents region has been built on the
premise of environmental cooperation by working toward a more sustainable region.
However, considerable transformations over the last two decades, such as large
investments and changing political situations, have altered this structure in some
ways to accommodate other areas of cooperation. Moreover, there have been political and economic changes that sometimes were out of the region's control and have
led to profound effects on the environment. The effects of the environment's impact
on food security through climate change and human activities are described below.

6.2 Climate Change

Climate change and environmental degradation can lead to increased temperatures.
For example, between 2004 and 2014, the Barents region topped the global average
temperature change of the last century (Vassilieva 2015). Northern Sweden had the
largest temperature increase in the region, as the average annual air temperature
went up by 1.4 °C. The county of Västerbotten had a particularly big jump of (+3.3)
degrees from 1.9 to 5.2 °C (Vassilieva 2015). Furthermore, the average annual temperature in Norway was 2.2° higher than normal. It is believed that Finland's average annual temperature has risen faster than anywhere else in the world (Vassilieva
2015). For the five Barents sub-regions in Northwestern Russia, the average annual
air temperature rise was 1.3 °C, but there were considerable variations between the
sub-regions; for example, the Komi Republic saw a 0.5 °C increase, while the
Republic of Karelia had an increase of 2.7 °C (Vassilieva 2015).

The problem of global warming is more pronounced in northern latitudes, especially in the Arctic-Barents region, than in any other region in the world. When there
is an increase in ice or snow cover on the Earth's land and seas, there is an increase
in the Earth's albedo—that is, the fraction of incident solar radiation reflected back
to space (Coakley 2003). As a result of global warming, the effects of "albedo," or
the reflectivity of snow, decreases, causing more heat to be absorbed, which leads to

an increase in surface and air temperatures, thus further hastening the thawing of snow in the Euro-Arctic-Barents region. One critical consequence of this is the limited access to game hunting, since the means of transport through the use of snow mobiles will be restricted. The ecological implications of a warming Arctic are mixed—some species will benefit, and others will be threatened. The positive feedback mechanism associated with the microbial conversion of organic matter to carbon dioxide and methane in thawing permafrost is known to enhance warming. On the other hand, the process is also associated with the release of nitrates that stimulate plant growth and atmospheric carbon dioxide uptake (Incopera 2016). The outcomes of these changes are becoming more evident through the environment and ecology of the Barents region, such as the changing vegetation and wildlife on land, something that is already noticeable in the area.

The Barents region, as previously described, is a region that expresses variety in food, people and land. These characteristics are being impacted by climate change as well as human activities, such as mining, oil and gas extraction, forestry, shipping and tourism. As a result, pollution and cross-border pollution has had outstanding effects on the northern communities across the Arctic and in the Barents region. Climate change is one of the most detrimental threats that we face today. Climate change was described as a real and significant threat not only to food security in the Arctic-Barents but also to the existence of northern indigenous peoples (Paci et al. 2004). A recent study by Henson and co-authors used computer models to examine how oceans would cope over the next century under a "business as usual trajectory" and a more moderate scenario in which the mitigation efforts that were promised under the Paris agreement come into effect. They showed that large swaths of the ocean will be altered by climate change (Henson et al. 2017). There will be rising temperatures, ocean acidification and lower oxygen levels in more than half of the world's oceans, which will make some organisms unable to tolerate the changed conditions, and they may be forced to migrate, evolve as a species or face possible extinction—leading to decreasing food supplies (Henson et al. 2017).

The repercussions of climate change have drastic effects on the four pillars of food security: availability, accessibility, utilization and food systems stability (FAO 2016). The availability of food is determined by the physical quantities of food that are produced, stored, processed, distributed and exchanged. Physical quantities of food in the Arctic-Barents are constantly fluctuating with shifting seasons, temperatures and weather patterns. These variations cause unpredictability in regard to how much or when most food will be available. Traditional food from the ground, such as berries, shrubs and vegetables, are subject to such alterations (Paci et al. 2004). The hunting of animals can become a daunting task; migration patterns adjust, making it difficult to track caribou, bowhead whale, fish and certain bird species. Therefore, due to climate change, the availability of such food sources has become unpredictable and unreliable, in which case the trends in high market prices of food around the world are usually a reflection and determinant of inadequate availability—and now these impacts are being felt in the Arctic-Barents (Paci et al. 2004). The FAO stated that when the prices of food are high, poor people are forced to reduce their consumption below the minimum required for a healthy and active life, which may lead to food riots and social unrest (FAO 2008); in other words, it may be said that "a hungry person is an angry person." Therefore, as the availability of traditional foods becomes unpredictable, the dependence on store-bought food will increase in the region.

Accessibility, as defined by the FAO, is a measure of the ability to secure entitlements, which are those sets of resources (legal, political, economic and social) that an individual requires to obtain access to food (Sen 1989). The definition has since expanded, now recognizing both the individual and household access to food (Pinstrup-Andersen 2009). The environment is a crucial resource for obtaining access to food in the Arctic-Barents, where changes to the environment could pose challenges to food accessibility. The supply of meat from reindeer as food is threatened when reindeer herders, who migrate with their reindeer during the summer and winter months, cannot find appropriate feeding grounds for the herd to feed them sufficiently. This has been attributed to changing weather patterns, and the frequent thawing and freezing has hampered the reindeers' accessibility to food under the snow (Turunen et al. 2016). In addition, access to other traditional foods such as grouse, elk, fresh and saltwater fish and berries that the Saami rely on is also changing with climate patterns (Berg 2014).

Food utilization refers to the appropriate nutritional content of the food and the ability of the body to use it effectively—in other words, the safety and social value of food (Burke and Lobell 2010). The nutritional content that the body needs to survive, such as protein, vitamins, oils, natural fats and other nutrients, is present in the traditional foods commonly consumed by the indigenous peoples of the region. A lack of this food due to limited accessibility and availability could promote the consumption of store-bought foods and lead to further health implications. Evidence indicates that warming temperatures in the Arctic-Barents can threaten the safety of food, as pathogens tend to thrive in this atmosphere (Burke and Lobell 2010). This presents food safety as a deep concern throughout the whole food systems process, from hunting the animal to storage, cooking and consumption (Burke and Lobell 2010). These processes are not in a controlled environment with packaging and storage facilities, so the risk of pathogens and bacteria in food can be significantly higher. As hunters have to travel further into the forest or to remote areas to hunt for game, the additional time it takes to transport it back to their community will result in an increase in bacteria. For example, due to the non-availability of cold storage, hunted game or fish will deteriorate faster over a longer period before they are able to be processed for food. Furthermore, many of these pathogens to which marine mammals are susceptible are of concern to human health, as they can be transferred between animals and humans (Burek et al. 2008).

Lastly, climate change can affect the overall stability of the food systems, which is when one or more of the four components of food security becomes uncertain or otherwise insecure. This food system stability refers to the overall balance of the food supply and is determined by the temporal availability of and access to food (FAO 2008). The globalized economy is highly sophisticated, but has managed to provide consumers with the availability of food and accessibility to food. However, predictions of weather changes and increased temperatures could threaten the current food systems that are in place, leading to astronomical ramifications (FAO 2008). In recent years, there have been surges of food riots all around the world in response to the increased price and limited availability of foods. The food systems in a given country are not dependent only on local changes—external forces also affect them, and we face these vulnerabilities on a global level. People in the Arctic-Barents are experiencing the high cost of groceries from imported foods as they move away from traditional foods into a more market-based economy (Nuttall et al. 2005). These changes have an impact on human health, livelihood assets and food production and distribution channels and also change purchasing power and market flows (FAO 2008). Numerous impacts overlap one another, creating domino effects in different areas of the food systems. This reveals the power that climate change has over the lives of the Arctic-Barents peoples. Furthermore, it is believed that if these challenges persist, the crises in food supply due to temporal fluctuations in food resources are expected to occur more often and be longer and more intense (Paci et al. 2004). In addition, the ecosystem services in the region will change, which will undermine the therapeutic benefits that humans obtain from ecosystems that are useful for their wellbeing. These ecosystems contribute to human wellbeing by providing culturally important species and scenic landscapes, and ecosystem changes may result in a loss of cultural ties to the land. Jansson et al. (2015) further explained this connection that ties to the land are partly manifested through the harvesting of products provided by ecosystems, allowing locals, including indige-nous peoples, to continue traditional land-use patterns. Such ties to the land may be affected by changes in the geographic distribution of species in response to climate change as well as by ongoing urbanization and globalization. The warming climate may introduce new host species and pathogens into new habitats that may lead to infections in humans. Increases in the morbidity rates of zoonotic infectious dis-eases (such as tick-borne encephalitis, tularemia, brucellosis, rabies and anthrax) among humans, domestic animals and wildlife in the Russian Arctic (Revich et al. 2012) and Alaska (Hueffer et al. 2013) have been reported.

The Barents Sea is one of the most dynamic and productive ecosystems in the world. It supports food webs that include large populations of seabirds, marine mammals and other species that are targeted by regional fisheries (Akvaplan-Niva 2007). The Barents Sea is home to more than 200 species of fish, including capelin, polar cod and juvenile herring, which are exploited commercially (AMAP 2017a, b). In the Barents Sea, the benthic organisms are dependent upon ice algae as well as phytoplankton as food. It is assumed that less ice algae and more phytoplankton will result in differences in food quality and can lead to risks for benthic organisms through changes in their allocation of energy to growth, reproduction and mainte-nance activities (Akvaplan-Niva 2007). Moreover, changes in the marine environ-ment could also see a large impact on the fishing industries as temperature changes in the water may cause fish to move northward, in which case fish from the South may swim into northern communities, like to Bodø, Norway, for example. This could have outcomes related to the diet of local peoples and their communities, with more fish and less meat (Van Oort et al. 2015). Others suggest a northerly migration of plants and animal species, as the tree line may move northward as well, which can be both good and bad for the region. The increase in species could replace those leaving the region, but a decrease or extinction of some species may occur as well in this process, where species may invade from the South (Eskeland and Flottorp 2006). Also, warmer weather may lead to an increase in forest pests such as pathogens, diseases and invasive herbivorous insects and plants (Neuvonen and Viiri 2017). In addition, berry production could see a decline, and some berry plants might be less common or non-existent (Wartiainen 2007). Alternatively, while warmer weather has a negative effect in most places, it could also enhance the food security of some communities with the introduction of newer species, provided that the community knows how to utilize these species (Eskeland and Flottorp 2006).

Reindeer husbandry could see changes, as this traditional activity is highly dependent upon the environment for grazing land and calving grounds. It was suggested that herding strategies are also shaped by factors such as season, snow type, temperature, landscape, unusual weather conditions and the physical condition of the animals (Magga 2006). When these conditions are continually changing, herding becomes highly unpredictable and unreliable, especially when moving reindeer by foot because of melting ice and snow conditions. This is of particular concern in Russia, where the herds have to be monitored closely by the herders due to low state subsidy (Rees et al. 2008). Furthermore, many have noticed that autumn and winter rain-on-snow (ROS) events, resulting in ice-encrusted pastures and the resultant mass starvation of semi-domesticated reindeer (*Rangifer tarandus*), have increased in frequency and intensity across the northwest Russian Arctic (Bartsch et al. 2010). This region is home to the world's largest and most productive reindeer herds. Warmer/wetter winters have also negatively affected the much smaller wild reindeer populations in the High Arctic Svalbard (Hansen et al. 2014).

These expectations can be the result of the Barents region experiencing higher temperatures and precipitation and less snow cover over the last decades (Serreze et al. 2000; ACIA 2005; IPCC 2013). Such changes, coupled with threats of depopulation and globalization, can be quite severe, and many herders risk the starvation of their herds. In addition, some expect that by changing harvest methods, local communities, and often entire societies, run the risk of losing local and traditional knowledge (Heleniak 1999).

Climate change in the Barents region poses both challenges and opportunities. Challenges to infrastructure with decreasing permafrost and difficult sea conditions for shipping and off-shore oil installations are a few of the concerns raised. On the other hand, there will be opportunities for the oil and gas industry, with a greater exploration area and for a longer season; for the fisheries industry, with greater fish stocks; and for the tourism industry, with a longer ice-free tourism season (Rottem and Moe 2007). Natural resource sectors, such as forestry, mining and oil and gas extraction, are important to national as well as local and regional economies (Glomsrød et al. 2009). However, these opportunities do not benefit everyone, and some will feel the negative effects of such industrial activities. In addition, ice-free tourism is not beneficial for everyone—northern Finland thrives immensely off of winter tourism, and changes to this would drastically impact the tourism industry there.

The increased green focus in transport, carbon footprints, continued local fisheries and changes in diet would probably make the region more self-sufficient in terms of food production, as increased temperatures allow for opportunities to cultivate a larger variety of crops in the region (Van Oort et al. 2015). However, persistent climatic variability resulting from ongoing climate change will bring uncertainty. As a result, major resources are devoted to preventing various impacts of climate change, such as flooding. Additionally, Barents Watch (2007) described some of the negative changes that could take place in the region, including that there will be a northerly migration of plant and animal species, some tundra areas will disappear from the mainland and the tree line will move north and increase in altitude in the mountains (Barents Watch 2007). Increased temperature will probably result in more damage to the forest as a consequence of insect attacks, and the overgrowth of open landscapes (the tundra) will decrease the nesting areas of many birds and the grazing land of many land animals. It is expected that rare animals may be lost and that species available at present may decrease considerably. Many of the animal species adaptive to a cold climate might be replaced by species migrating northward because of the warmer climate (Hossain 2015a, b, c). For some bird, fish and butterfly species such displacement is already under way, and a warmer and more humid climate might also result in reduced berry production, and berry plants might become less common (Wartiainen 2007).

It is likely that the consequences of climate change to the Barents region outweigh the benefits of such a phenomenon. Many wildlife species previously used as food sources have disappeared due to climate change. In addition, the reduction in snow-cover in winter has affected hunting, travel and other transportation. Climate change is having a wide range of impacts on animal and plant species, landscapes and other environmental ecosystems that are important to the people of this region. Changing such landscapes can reduce the environmental security of the individual and the community, which is a key part of individual and collective human security.

6.3   Human Activities

The European North is rich in renewable and non-renewable natural resources, such as forests, fish, minerals and fossil fuels. As climate change becomes more pressing in the Barents region and on those living there, it also opens doors and opportunities for increased human activity, as discussed above. A decline in summer ice will open the door to commercial opportunities with greater access to large fishing grounds, the expedited transport of goods and off-shore exploration for oil and gas (Krauss et al. 2005). Increased interest in these resources will grow as the region becomes more accessible and the technologies to operate in such harsh northern conditions are developed. There are concerns that these developments will intensify the conflict between man and nature as well as conflicts between different economic activities— for example, traditional livelihoods and eco-tourism versus extractive industries (Coria and Calfucuria 2012). As more investment goes into the region, it will attract more people and increase human activities, and these human activities are increasingly threatening to food and human security. As the community becomes more urbanized due to increased population, there will be a need to provide infrastructures and health services that will minimize the spread of infectious diseases. An increase in infections such as gastroenteritis, respiratory infections and vector-borne diseases will lead to decreased access to safe food and water (AHDR II 2015). The One Health Initiative concept recognizes the fact that the health of people is connected to the health of animals and the environment. Therefore, there will be a need to work with physicians, ecologists and veterinarians to monitor and control public health threats. This is carried out by learning about how diseases spread among people, animals and the environment (One Health Initiative 2016). These new pathogens and pests with the adverse health impacts of Arctic warming will have negative impacts on wildlife populations and indigenous peoples, who are dependent upon subsistence food resources from wild plants and animals (One Health Initiative 2016). However, there are also some positive effects of Arctic warming in terms of food security in the circumpolar North—for example, an increasing potential for agriculture during the longer growing seasons (AMAP 2011a, b).

Food costs in the Arctic-Barents regions are higher in comparison to the southern counties. In the Russian Barents, for example, 23–43% of household income is spent on food (Dudarev et al. 2013). There have also been high levels of biological and chemical contamination of food in many regions (Dudarev et al. 2013). In a study by Dudarev et al. (2015) on the toxic metal levels in local food items like fish, mushrooms, berries and game in the Pechenga district, the researchers found high cadmium, nickel and copper concentrations in mushrooms, and high nickel levels in wild berries. All these human activities can cause tremendous impacts on the food security situation in the Barents region through impacts on the environment, health and community.

6.3.1   Mining

Mining is a large industry that is present across Norway, Sweden, Finland and Russia, extracting gold, copper, silver, iron, zinc, phosphorus, apatite, nickel and other elements from the ground (Eilu 2012). Mining is a deep concern for people living in the vicinity of these mining sites, as such activity has previously caused damage to forests, traditional hunting grounds and waterways. Greenpeace believes that strip mining can destroy the landscapes, forest and wildlife habitats around the mine site, where rain takes the loose top-soil and can wash it into waterways, hurting fish and other wildlife and can lead to the chemical contamination of the groundwater as well as noise pollution and dust from machinery (Greenpeace 2010). Therefore, the consequences of such mining actions can have dangerous impacts on traditional food sources in the Arctic-Barents region (Herrmann et al. 2014). Industrial activity and motorized vehicles pose a hazard to nearby animals and species, leading to a loss of wildlife habitat (White et al. 2007). Animals and humans alike are easily exposed to the digestion of chemicals and heavy metals, which are often found in the water supply (Tchounwou et al. 2012). Noise from mining activities can be potentially disruptive to wildlife, causing nearby species to migrate further, forcing hunters to travel a greater distance or find new hunting grounds all together (Shannon et al. 2016).

The Russian border town of Nikel, located in Murmansk Oblast, provides a good example of such disruptive industry, as its pollution has been at the top of the environmental agenda of the Barents Cooperation (Nilsen 2016). In the early 1990s, the plant in Nikel emitted up to 300,000 tons of sulfur dioxide per year. Today, the emissions are significantly lower, but the plant remains by far the worst polluter in the region. Figures from the Federal Hydrometeorology Authority in Murmansk, for example, show that the plant on August 2, 2016, emitted up to eight times the allowed level of sulfur dioxide (Staalesen 2016). Norilsk Nickel, the mother company, also runs a string of other heavily dirty production facilities in both the Kola Peninsula and the west Siberian Taymyr Peninsula, making it one of the worst polluters in the whole Arctic (Staalesen 2016). Furthermore, the Kola Company is spread out across Northwest Russia in three of the dirtiest industrial towns in the country: Nikel, Zapolyarny and Monchegorsk (Kireeva 2016). The pollution from these sites has been known to have drastic effects on nearby local wildlife, the environment and even food sources, such as berries and mushrooms (Hansen 2016). In the Scandinavian North there is continually increasing mining activity, and some say it will bring permanent damage to the vast network of rivers, lakes and mountains, which are home to many of Europe's largest mammals, such as the lynx, wolf, bear and wolverine (Vidal 2016). Furthermore, some human rights groups argue that Lapland and the Saami indigenous communities who make a living from reindeer herding and fishing will be badly affected by the development of the region's tourist industry, which depends on pristine nature (Vidal 2016).

6.3.2 Oil and Gas

The oil and gas industries have long been established in the region, where Norway and Russia are world leaders. These countries are also significant exporters of both products (Austvik and Moe 2016). Russia intends to diversify its oil exports and oil transport in the area, which is also expected to grow significantly over the coming years; significant volumes of oil products are also being shipped from various harbors in the Russia Barents region, including Murmansk, Vitino and Arkhangelsk (Austvik and Moe 2016). Norway started exploration in 1980 at the Tromsøflaket offshore bank in the southwestern part of the Barents Sea, and there has been several minor and two major discoveries since then—the Snøhvit gas field, discovered in 1984, and the Goliat oil field, discovered in 2000 (Moe 2010). There are worries about these operations due to the harsh climatic conditions prevailing in the Barents Sea and problems that might be related to cleaning oil spillage in the sea. The World Wide Fund for Nature (WWF), has expressed concerns over ineffective clean up methods in instances when spills do occur (WWF 2016). There are also concerns about the noise generated, which can injure marine animals, as these species use sound to navigate the ocean to find food, and both drilling and production can disturb the fish and other animals, which have both economic and food value (WWF 2016) for humans to consume. Mark Nuttall (2000a, b) explained that polar bears, seals, sea otters and sea birds are already frequent casualties of oil contamination, while bowhead whale migration routes through oil and gas lease areas could be seriously disrupted if development goes ahead. It was also suggested that a single serious oil spill could destroy entire populations and greatly endanger unique species, particularly should the event overlap with the presence of migratory species, which often congregate in relatively small areas (Hossain et al. 2014). Moreover, such impacts can affect the food situation of larger animals and nearby communities and populations who rely on these species. However, in recent years the Barents Sea has seen relatively slow development due to the costs and political risks involved in addition to being hampered by recently introduced sanctions and a slump in the price of oil (Kullerud and Ræstad 2016). Accordingly, such developments may not proceed for a while.

6.3.3   Forestry

Forestry is a prominent industry in all the four countries of the Barents region. Alterations in forest use management result in impacts on traditional food system in various ways. In Finland and Sweden, forests comprise 74% and 60% of the total land area, respectively (Baldursson 2003), and in Russia it is 49.8% (World Bank 2015). In Finland, according to the FAO, forestry accounts for 4% of the gross domestic product (GDP) and is the highest among the developed countries in Europe (Lebedys and Li 2014). In these four countries, after decades of sustainable practice, forestry has become much more sustainable. However, despite the increased management of sustainable forestry resources, there is no denying that it has had considerable influence on the local environment. The use of the land and environment is common to both indigenous and non-indigenous local populations, who often rely on forest resources, such as for their traditional hunting, gathering and cultural practices. This has resulted in a number of disagreements between the forest industry and the local indigenous and non-indigenous peoples over the effects of forestry-related activities on their livelihoods (Lawrence and Raitio 2006). Among the many issues that surround cutting down forests, one is their use by reindeer for food such as lichen that grow on the ground and the tree-hanging lichen. During the winter months, reindeer are herded to specific locations in the forest to graze. These herding areas consist of coniferous forests, which are heavily exploited by the forestry industry (Lawrence and Raitio 2006). Many companies, such as those in the forestry and mining industries, focus primarily on commercial objectives and may or may not realize the impacts they have on the way of life of many northern communities (AWRH 2016). The forests are not only used for herding; all across the

Barents region they contain a great source of food, including traditional foods such as berries, herbs, shrubs, lichen, moose, caribou, lynx and wolverine.

6.3.4 Shipping

Shipping has increased drastically in the coastal areas of the Arctic-Barents region, as the sea ice continues to retreat progressively each year (Liu and Hossain 2017). Murmansk, Russia, has a year-round ice-free port. The city of Murmansk is regarded as a prominent shipping hub within the region. The Northern Sea Route Information Office has started recording the number of transits across the Northern Sea Route; in 2011, they recorded 41 vessels, while in 2013 it reached a peak of 71 (McDonald-Gibson 2013; Liu and Hossain 2017). The Northern Sea Route stretches across Russia's northern coastline and connects Europe to Asia, and it is believed to hold substantial time and cost savings for international shipping companies. These routes are estimated to save hundreds of thousands of dollars each year through cost and time savings (Rahman et al. 2014). It was estimated that ships can save about 12 days in their journey by using the Northern Sea Route and save both fuel and money (McDonald-Gibson 2013). However, the coastal areas, especially those around the Barents Sea, are used for fishing, hunting and other economic benefits; these could be affected by a surge in Arctic-Barents shipping.

The increase in maritime activities in the region is expected to have grave consequences on the marine environment, as disruptions to the natural mating and migration patterns of fish could occur. In turn, humans and animals, who rely on the marine ecosystem for food and traditional practices, will be at a loss (WHO 2016). Further, as a result of increased traffic in maritime activities, there could be more discharges from marine vessels—such as tankers, freighters, fishing boats and coastal ferries—which may not be readily monitored, but their impression on the Arctic-Barents ecosystems will be substantial nonetheless (Nuttall 2000a, b). These discharges refer to the ballast water that ships use to balance the load of the ship. This water is often gathered from the South and dispersed in the North, with the potential to release invasive species (AMSA 2009; EMSA 2014; Hossain 2015a, b, c). Although the outcomes of invasive species are still being researched, Lisa Palmer, the author of "Hot, Hungry Planet: The Fight to Stop a Global Food Crisis in the Face of Climate Change" published in 2017, believes that shipping is by far the most common pathway for marine invasive species, responsible for 69% of species introductions into marine areas (Palmer 2013). The effects of such invasive species still need to be researched to a greater degree. However, according to Lassuy and Lewis (2013) the potential pressure on biodiversity resulting from the introduction of invasive species contributes to biodiversity loss, which eventually affects humans' ability to use that biodiversity. Moreover, there is always the risk of toxic chemicals and oils that could be accidentally released into the environment, which is considered one of the most serious threats posed by shipping (Ellis and Brigham 2009). Therefore, the result of increased shipping, and resulting consequences, in particular the introduction of invasive species, have detrimental effects on animal populations and eventually on the food security of nearby populations. Robards (2013) remarked that without policies that proactively address the risks associated with large vessels transiting in marine mammal hotspot areas, or areas that support indigenous subsistence practices, negative impacts on these populations and indigenous food insecurity will be on the rise.

6.3.5 Tourism

Tourism is a growing industry in the North, as people generally aim to explore the relatively unexplored region. The uniqueness of the Barents region attracts tourists to travel this wonderfully pristine region. However, similar to the increase in activities related to shipping and ferry services, ship traffic carrying tourists is likely to carry risks of vessel accidents, oil spills and pollution from discharges, resulting in devastating effects on the pristine Arctic environment (AMSA 2009). The Barents region has developed tourism strategies on levels ranging from national to regional themes. The natural and geographical conditions that are common in the region attract tourists for winter activities—for example, the northern lights phenomenon and midnight sun in summer. The state of the tourism industry varies greatly among the Barents countries based on the varying numbers of turnover and employed workforce it creates (Kohllechner-Autto 2011).

According to an analysis of the strategic development of tourism in the Barents region conducted by Kohllechner-Autto (2011), tourism was shown to be well developed in Norway, Sweden and Finland. In these countries, tourism is based on a well-established infrastructure, and these destinations are well known both nationally and internationally, while in the Murmansk and Arkhangelsk regions of Russia tourism is still in its infancy.

Table 6.1 shows some indicators of tourism in the Barents region. The table compares the contribution of tourism as an industry and as a contributor to the economy of selected counties and cities in the region. The statistical figures are from 2009. The number of visitors is on the rise—especially from Asian countries—a trend that is on the upswing in the Barents region. However, regardless of the different stages of tourism development within the Barents region, certain common challenges can be observed, especially in Finland and Sweden, resulting in the strong dominance of winter tourism with a relatively weak summer season. This is a challenge in terms of year-round employment for the population of the region; on the other hand, it will be an added advantage to encourage food business operators, who can add value to the traditional foods and deliver year-round employment opportunities.

Tourism in Northern Norway is similar to tourism in Finnish and Swedish Lapland, as it relies heavily on its natural beauty and climate in attracting tourists. The tourism infrastructure in Norway is characterized by the prevalence of large industrialized hotel chains against a cluster of small-scale enterprises that provide food, accommodation and various experiences (Kohllechner-Autto 2011). The interest shown by

Table 6.1 Some indicators of tourism in the Arctic-Barents region[a]

Region, Country	Indicator	Value
Finnish Lapland, FINLAND	Tourism industry units	1200
	Tourism industry turnover/revenue	€ 540 mil.
	Direct annual tourism industry income	€ 594 mil.
	Total number of annual tourist arrivals. Registered tourist overnights (foreign, domestic)	950,000 240,000
Swedish Lapland, SWEDEN	Tourism industry units	550
	Tourism industry turnover/revenue Tourism industry man-years	SEK 3.37 bil. 2500
	Registered tourist overnights	2.3 mil.
Murmansk, NW RUSSIA	N/A	
Monchegorsk, NW RUSSIA	Tourism industry business units	81
	Tourism industry man-years	508
	Total number of annual tourist arrivals	1525
	Registered tourist overnights	9498
Arkhangelsk, NW RUSSIA	Tourism industry business units	635
	Tourism industry turnover/revenue	RUB 1046.7 mil.

[a]It was not possible to obtain data for Norwegian Lapland; the only available figures were for the entire country. Figures were not available for Murmansk, either. 'Man-years' refers to the average hours an individual works in a year. (Adapted from Kohllechner-Autto 2011)

tourists in different food cultures has been described as a factor for local development in the fields of agro-food and crafts whilst also contributing to the enhancement of food culture and heritage (Bessiere and Tibere 2013). The Norwegian government's vision for its tourism strategy is similar to that of Finnish Lapland in the sense that it aims to provide valuable and meaningful experiences for tourists (Kohllechner-Autto 2011). The three main objectives of the government's tourism strategy are (1) to achieve greater wealth creation and productivity in the tourism industry, (2) to create sustainable rural communities through year-round employment in tourism and (3) to build up Norway as a sustainable destination (Kohllechner-Autto 2011). Tourism in the Murmansk area is still in its early phases; the city attracts mostly business tourists and a few international leisure tourists. However, Kohllechner-Autto (2011) stressed that many different tourism projects are currently being planned in the Murmansk region, such as several tourism and recreational territories (e.g., in Kirovks and Lovozero).

In the Barents region, tourism is a prominent industry, where visitors come from all over the world to see the northern lights, reindeer, Santa Claus and other Arctic-specific adventures. Although tourism brings revenue, which would boost the region's economy in the future, it is expected to offer new and real threats in the Arctic-Barents, in particular concerning the region's natural environment. For instance, the use of motorized vehicles such as helicopters, snowmobiles and planes for recreational purposes can be disturbing wild animals. The use of these vehicles also generates environmentally detrimental carbon footprints, and they create noise that forces some animals to relocate. When helicopters and airplanes are used to convey tourists frequently into the region, the noise they produce can cause panic flights of seabirds and other birds, which will eventually lead to a reduction in their eggs (Snyder 2007). In addition, sports fishing and hunting has gained popularity in the Arctic-Barents, but they frequently put pressure on resources, leading to conflicts between local and visiting hunters as well as among the different interest groups (Snyder 2007). Overall, a surge in tourism means a rise in garbage and waste, exclusively in an area where decomposition is slow, and waste remains visible on top of the permafrost in many areas (Huntington et al. 2001).

Apart from the detrimental effects of climate change and human activities on the environmental security of the northern areas, a warmer climate will encourage new economic activities that will further put pressure on natural ecosystems and the people who are dependent on the functioning of these ecosystems. Because of the increased occurrence of extreme weather and the melting of permafrost there will be new challenges to manmade infrastructures as well as to natural systems. It was suggested that, when considering new developments, not only should the loss of biodiversity and human health be seriously weighed, but consideration should be given to alternative ways of using these areas that will promote long-term environmental and economic security (FCES 2014). In terms of possible alternatives, the potential of renewable energy resources and sustainable nature tourism as sources of employment are areas that remain to be explored.

Overall, tourism is an important industry that helps to bridge cultural differences, and by nature humans are curious to learn about other cultures, which includes food and traditions. The curiosity about emblematic products and foodstuffs from the regions visited is an important driving force in the tourist experience, along with interest in natural sites or in architecture and famous monuments (Bessiere and Tibere 2013). Food tourism can serve as a vehicle for regional development by strengthening local production (Renko et al. 2010). The traditional and local foods in the region when purchased by tourists will help to support the small and medium enterprises that are sellers of such foods, and this will further promote the cultural identity of the region.

6.4 Impacts of Contaminants in the Food Supply System with Special Reference to Barents

Due to climate change and increased human activity, contaminants and pollutants are entering the Arctic at an increasingly alarming rate (Nuttall 2000a, b). Contaminants, introduced via direct exposure to pollution accruing because of environmental change as well as via the transportation of long-range atmospheric currents into the marine environment and freshwater/terrestrial routes can have a severe impact on the region's environment, affecting the health of the animals and populations living there (AMAP 1998). These contaminants have the potential to bio-accumulate in the food chain on which individuals and communities rely. AMAP emphasizes that exposure to environmental contaminants through the traditional diet remains one of the greatest risks to human health in the Arctic (AMAP 2015). Indigenous peoples in the Arctic often consume a great deal of these food resources, and they have shown higher levels of organochlorines in blood and breast milk than in those of the non-indigenous populations living in more urban environments (Kuhnlein and Chan 2000). This is particularly true for the population living in the Canadian North (Kuhnlein and Chan 2000). However, this is not the case in the context of the Barents region, since it consists of both indigenous and non-indigenous peoples, who are quite often mixed in many communities. They both use the same environment as a means for food.

The contaminants can come from local sources during mining, which produces heavy metals such as mercury and arsenic, or from abandoned military sites, which may entail PCBs, pesticide use, dichlorodiphenyltrichloroethane (DDT) and toxaphene (Muir et al. 1992; Volder and Li 1995). Other sources are radionuclide contaminants mainly from nuclear testing, nuclear power stations or nuclear satellites (Barrie et al. 1992). Pollutants are detected in the air, fresh water, seawater, snow, sediments, birds, fish, plants and terrestrial and sea mammals, where they can be accumulated in food species all across the Arctic (Barrie et al. 1992; Muir D. et al. 1999). The Arctic-Barents region has been described as a sink for global pollutants despite the fact that the region is distantly located from heavy industrialized centers and agricultural source regions (AMAP 2016). Pollutants are released at lower latitudes and are deposited in the Arctic ecosystems by the atmosphere, oceans and rivers. There are concerns that the current global regulatory practices do not account for the emerging pollution threats that are not persistent organic pollutants (POPs). For example, "microplastics" are emerging as a major environmental concern worldwide, including in the Arctic-Barents. Microplastics exhibit some similarities to POPs in terms of their long-range transport and their potential for causing harmful effects. However, due to their complex makeup, they cannot be evaluated with the current risk assessment tools and criteria used for POPs, which were developed to focus very specifically on individual chemicals with specific properties (AMAP 2016).

Plastic has a large impact on the environment and ecosystems; it is estimated that by 2025, the oceans will contain one ton of plastic for every three tons of fish, and by 2050, the oceans will contain more plastic by weight than fish (Macarthur 2016). Within the last 10 years, more plastic has been produced in the world than in the entire twentieth century, and, notably, plastic causes more than USD 13 billion of damage to marine ecosystems every year (UNEP 2014). In response to this dilemma, a new endeavor, "The Seabin Project," in collaboration with the Finnish Wärtsilä Corporation, aims to provide solutions to the problem of littering in seas around the world. The project approaches the challenge from several angles, with special focuses on education, research and technology (Wärtsila 2017). The "Seabin" is a floating rubbish bin that is located in the water at marinas, docks, yacht clubs and commercial ports, where it collects all floating rubbish (Klein 2016). Water is sucked in from the surface and passes through the catch bag inside the Seabin. The water is then pumped back into the marina, leaving the litter and debris trapped in the catch bag to be disposed of properly. The Sea beans also has the potential to collect some of the oils and pollutants floating on the water's surface, which will help to save the fishery and aquaculture resources in the Barents Sea. The Seabin Project's team currently uses submersible water pumps that can utilize alternative and clean energy sources, including solar, wave or wind power, depending on the location and available technology (Wärtsilä 2017). Van Sebille and colleagues predicted that most of the buoyant plastics thrown into the sea from lower latitudes creates a plastic accumulation zone within the Arctic Polar Circle, specifically in the Barents Sea (Van Sebille et al. 2012). This was further confirmed in a recent study showing that, despite the fact that plastic debris has historically been scarce or absent in most of the Arctic waters, it has recently reached high concentrations of up to hundreds of thousands of pieces per km^2 (Cózar et al. 2017). An efficient means of collecting plastic waste, such as the Seabin, will be helpful for the coastal communities of the Barents region, since they rely on traditional aquatic and marine animals.

The Barents Sea is the largest among the pan-Arctic shelf seas that surround the Arctic Ocean (AO), covering about 30% of one of the world's largest shelf sea expanses, linked through the inflows of Pacific Water (PW) and Atlantic Water (AW) (Ingvaldsen et al. 2002; Schauer et al. 2002a). The Barents sea is one of the world's largest fisheries; the permanently ice-free waters in the South and Southwest, close to no freshwater supply by the rivers in its central and northern regions and flow-through of significant fractions of AW or locally modified AW (Schauer et al. 2002a). The inflow from the Norwegian Sea is an order of magnitude greater than the inflow through the Bering Strait (Carmack et al. 2006), and this water flows either through the Barents or to the west of the Barents, entering the AO north of Svalbard (Wassman et al. 2006). The Barents Sea plays a crucial role in Norwegian and Russian fisheries and aquaculture, and it is also well known for its highly productive continental shelf sea area, which feeds many species of fish, seabirds and marine mammals. However, a wide range of chemicals that contain carbon, chlorine and, sometimes, several other elements known as organochlorines find their way into the sea and are a source of contamination in these species (Van Osstdam et al. 2005). A range of these organochlorine compounds have been produced, including many herbicides, insecticides and fungicides as well as industrial chemicals such as PCBs. In the western Barents Sea, many seabirds breed in Bjornoya, where a high concentration of organochlorines was found in the marine predator glaucus gill (*Larus hyperboreus*). Similarly, in the Svalbard archipelago located in the northwestern Barents Sea, high organochlorine levels were also found in marine top-predators such as the arctic fox (*Alopex lagopus*) and polar bear (Norstrom et al. 1998; Borgå et al. 2000).

Organochlorine compounds are potential endocrine-disrupting compounds that can interfere with the endocrine system and disrupt the hormone balance of an animal. This typically results in a disruption of the reproductive processes in aquatic organisms or causes immunodeficiency. Furthermore, endocrine-disrupting chemicals can mimic, interfere with or block the function of endogenous hormones and cause adverse developmental, reproductive, neurological, cardiovascular, metabolic and immune effects in humans (AMAP 2015). These effects can be caused by very low concentrations, considerably lower than those that can cause changes to genetic material (i.e., mutagenetic or acutely toxic) (OSPAR 2000). Organochlorine compounds are more soluble in fat than in water—they are lipophylic—which gives them a high tendency to accumulate in the food chain (biomagnification) (Gray 2002). Arctic-Barents organisms are highly dependent on lipids for energy storage. The predators in the Arctic marine food web are exposed to high organochlorine biomagnification due to rapid and efficient energy transfer coupled with a high lipid content (Norstrom et al. 1998). Due to the fact that organochlorines are stored in the fatty tissues of the body, they will only become biologically available and have effects when fat tissues are metabolized. Therefore, it is possible for animals to have a considerable body-burden of organochlorines, but these may only start showing effects in conditions of starvation, once the fat reserves are mobilized (De Pooter 2013). Another worrying factor is inadequate wastewater treatment, which seems to be a source of certain pharmaceuticals and chemicals that are used in personal care products as well as other chemicals found in household products such as siloxanes and phthalates (Arctic Now 2017).

Traditional food systems in northern Europe have shown high levels of lead in plant ash, fish, sea mammals and birds (Kuhnlein and Chan 2000). The sources of lead can vary from natural soil and water levels to lead shot used for hunting (Grandjean 1992). Cadmium sources in traditional food systems are highest in the organs of large mammals (e.g., caribou and moose) and fish (Berti et al. 1998; Kim 1998) but are much less significant than cadmium exposure from tobacco smoking (Archibald and Kosatsky 1991). These pollutants affect both animals and humans. Humans are exposed to these contaminants through their consumption of wild food, where certain environmental pollutants can adversely affect the development of the immune system (AMAP 2015). For example, the kidney is believed to be the target organ for cadmium toxicity, with bone disorders as a possible consequence of kidney malfunction; cadmium is also a lung carcinogen (Hartwig 2013). Mercury is another pollutant that can be disruptive in the human body, where long-term exposure can permanently damage the brain, kidney and developing fetus (Kuhnlein and Chan 2000). Furthermore, chronic exposure to arsenic can be carcinogenic and may lead to neurotoxicity, vascular disease and liver injury. Lead toxicity is especially dangerous for children, with effects detected in the nervous system, in blood cells as anemia and in damage to the kidneys (Goyer 1996). However, despite the known effects of these contaminants, there are no reports of their environmental levels in traditional-food-system species having resulted in mortality or morbidity (Van Oostdam et al. 1999).

A related endeavor studied the industrial impact of contaminants on food safety and human health in the KOLARCTIC KO 467 project (Heimstad and Sandanger 2013). The project investigated several contaminants and their impacts on local berries, mushrooms, fish, reindeer and moose in the Barents region of Norway, Finland and Russia from 2007 to 2013; an overview of the result is shown in Fig. 6.1.

The findings showed that the level of contamination of copper, nickel, arsenal, cadmium, lead, mercury, polychlorinated biphenyls, hexachlorobenzene, dioxins and furans ranged from low to moderate and were considered safe for consumption.

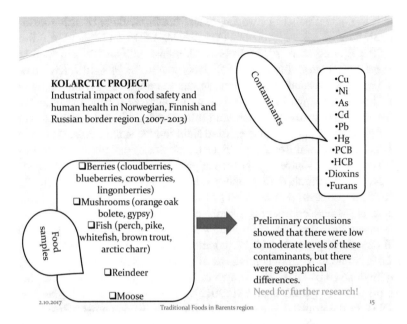

Fig. 6.1 Contaminants in foods in the Norwegian, Finnish and Russian border in Barents region. (Source: Raheem 2016)

However, there were geographical differences—the levels were higher along the Russian border—and there is a need for further research.

Although the research on contaminants and pollutants entering the Arctic has been primarily directed at the North American Arctic, there are some studies and reports, such as the AMAP 2015 report, that briefly looked at the European and Russian Arctic, too. In Norway, the levels of most POPs have declined significantly since 1979 in a single cohort of Norwegian men. This is consistent with the expected reduction in environmental exposure following international action across several decades to reduce or eliminate the production or use of POPs (Nøst et al. 2013). However, AMAP has documented that the consumption of marine mammals and fish was associated with increased levels of PFOS, PFNA, PFDA and PFUnDA and that beef consumption was significantly associated with increased levels of PFOA.

The consumption of game (e.g., reindeer, moose and grouse) was significantly associated with increased levels of PFHxS, PFHpS and PFNA (Nøst et al. 2013). In Sweden, the concentration of POPs in breast milk and its serum were found to be decreasing, with no strong indication of a generally higher concentration in northern Sweden in comparison to other parts of Sweden, which could be the general trend in the Arctic (AMAP 2015). This trend can be explained in relation to the highly centralized food distribution system in the Barents region nowadays.

The overall conclusion was that as long as store-bought foods make up the main proportion of the diet, there are only small or no regional differences in POP exposure at a population level (Glynn et al. 2011). Fish do remain a significant source of exposure to contaminants in Sweden, where the biomarkers of fish consumption correlate well with serum levels of PFOS, PFNA, PFDA and PFUnDA (Wennberg et al. 2007; Bjermo et al. 2013). In Finland, breast milk analyses in food sources from most store-bought foods have relatively low levels of PCBs (Kiviranta et al. 2001). There are elevated levels present in seafood originating from the Baltic Sea, but this is due to known PCB pollution in the Baltic Sea, which mainly concerns people consuming seafood in southern Finland and not in the North. In Russia, a study of male and female volunteers from Nelmin-Nos in the Nenets AO and Izhma and Usinsk in the Komi Republic indicated that older people in these northwestern Russian communities have significantly higher blood concentrations of many contaminants that accumulate in people over time, and higher PCBs and Pb were noticed in men compared to women (AMAP 2015).

Izhma is a small community mainly populated by reindeer herders, and the study showed that men have significantly higher concentrations of HCB, whereas overall the whole group has higher concentrations of cadmium (Cd). The group from Usinsk had higher concentrations of selenium (Se) and dichlorodiphenyldichloroethylene (p,p'-DDE, for women only), where most people make their living as oil and gas workers (AMAP 2015). In 2003, there were higher levels of total DDT in the non-indigenous population from Arkhangelsk, Russia, than in any other region, indicating a possible use of this pesticide locally or in Russian agricultural regions from which foods are transported to the Arkhangelsk region (Odland et al. 2003).

The consequences of activities from outside the region are making their way to the Barents region in the form of contaminants and pollutants through air and water currents. The highest levels of radioactive pollution along the Norwegian coast, for example, did not originate in Russia, however, but from radiochemical plants in the United Kingdom and France (Nuttall 2000a, b). Similarly, radioactivity from the nuclear test explosions carried out near Novaya Zemlya in the Russian Arctic affects the northern Atlantic and Barents Sea (Nuttall 2000a, b). These contaminants are having drastic outcomes on the health of the populations of the region in different variations and levels. Effects have been noticed in the human body such as in the kidney, the liver, brain development and others (Nuttall 2000a, b). There are indications that in the Nordic countries a large number of pollutants and contaminants are making their way into the fish and the marine environment, where humans are digesting these foods almost daily. On the other hand, store-bought foods have been associated with low levels of contaminants in these countries. The changing environment, coupled with contaminated traditional food, has led to an overall shift in diet in the Barents region and in the Arctic as a whole. This shift emphasizes the movement away from traditional foods, toward more store-bought "Western" or market foods, especially among indigenous peoples in the region (Duerden 2004). Regarding chemical contaminants in food, particularly fish, marine mammals, reindeer and caribou are of interest from an Arctic health perspective. However, it is important to monitor all food, not traditional food only (Nilsson and Evengård 2013).

6.5   Rise in Imported Foods

The phenomenon of an increase in imported foods has been taking place in the Arctic-Barents for decades, as individuals and communities switch from a traditional diet of fish, moose and reindeer to a Western-based diet consisting of—for example, chips, pizza and cookies imported to the region over long distances. There are large variations in the dietary patterns within the Euro-Arctic area, but there is a general tendency that traditional or country food consumption is gradually decreasing as imported foods are becoming more available and culturally acceptable to Arctic peoples (Odland et al. 2003). The AMAP 2009 report on health explained that diet is the single most important predictor of contaminant exposure in the Arctic-Barents populations (AMAP 2009). As discussed in Sect. 6.4 above, the traditional foods that people consume in the Barents region are affected by contaminants that are making their way into the region in one way or another. This poses a difficult daily decision for the Arctic peoples regarding whether or not to consume traditional food; choosing appropriate food can be a discouraging task. As AMAP (2009) described, traditional foods are an excellent source of nutrients and energy and contribute to good social, spiritual and physical health; however, they are also the primary source of POPs and metals. Therefore, there is a clear concern over contaminants, cultural values and the availability of traditionally hunted species due to climate change, which all play a role in influencing the types of traditional foods consumed, the frequency of their consumption and the exposure of Arctic populations to contaminants (AMAP 2009). Kuhnlein and Chan (2000) suggested that the continuing importance of traditional local food in the diets of indigenous peoples makes the issue of the contamination of these foods especially worrisome.

In contrast to market foods, traditional foods are related to cultural benefits, which include beliefs about food healthfulness and spiritual provisioning as well as the use of food for its educational value, economic benefits and place in the social fabric of community life (Usher 1995; Kuhnlein and Receveur 1996). The Saami diet prior to the 1800s, for example, was described as being much higher in meat and fish and much lower in cultivated vegetables and bread than today (Rheen 1983; Håglin 1999). Some describe this diet as that of stone-age hunter/gatherers or often referred to as a Paleolithic diet today (Haraldson 1962). In Russia in the 1920s, Saami people were estimated to eat about 70 kg of fish and 120 kg of reindeer meat per person per year (Kozlov 2008); in comparison, the average consumption in Sweden in 1990 was 30 kg of fish and 60 kg of meat per person (Becker 2008). The change in the amount of consumption over the 70-year timeframe may demonstrate how the diet has gradually changed or even perhaps diversified. The reasons for this change can be associated with the lack of or lower quantities of local traditional food resources and increased commercially produced market foods that are imported to the region (Kuhnlein and Chan 2000). The 2009 UNEP/Grid-Arendal report argued that this lower quantity is brought about by climate change, which poses a threat to traditional food security in northern regions because it influences the availability of animals, the ability of humans to access wildlife as well as the safety and quality of available wildlife for consumption (Meakin and Kurvits 2009).

The report cited examples of stronger winds making travel and hunting more difficult and dangerous by boat in the summer and the increased length of the ice-free season and decreased ice thickness making it more difficult and dangerous to access ice-dependent wildlife (Meakin and Kurvits 2009). AMAP described that for indigenous populations, by the turn of the millennium, most of their dietary energy was being obtained from imported food, and this pattern has remained the same or increased further. However, on the other hand, in Russia it was identified that the socio-economic changes and deterioration of the farming and livestock system in the northernmost parts after the dissolution of the former Soviet Union seem to have led to increased use of local foods in some populations (AMAP 2009). This is not the case for all populations, and it is likely not the case in Russia anymore, but it is believed that abandoning hunting for imported food would not only be less healthy but would also be immensely costly (Nuttall et al. 2005). The changes in food sources can be influential in terms of nutritional quality and density as well as food security for both indigenous and non-indigenous peoples (AMAP 2009). AMAP (2009) further explained that, nutritionally, the problem is not the imported food itself but rather the widespread replacement of traditional food by a diet that is high in sugar and foods with low nutrient density.

AMAP also pointed out that, in general, the decreasing proportion of traditional foods in the diet has had a negative impact on the intake of most nutrients, but some imported foods appear to have contributed positively to the intake of vitamin C, folate and possibly calcium (FAO 2013). According to the measurements from blood and dietary intake, nutritional deficiencies of vitamin A, iron, calcium and magnesium are prevalent in some communities due to a decline in the consumption of traditional foods (AMAP 2009). Therefore, the consumption of traditional foods, despite the concerns of contamination and pollution, are known to be beneficial for nutrition and health. There are also enormous health benefits associated with the traditional lifestyle—plant harvesting, hunting, fishing and berry picking require physical activity and experiences in nature (Kuhnlein and Chan 2000). Furthermore, when these activities become regular patterns, they contribute to overall fitness and general health. Store-bought or market-based foods do not hold the same nutritional value and health benefits. In fact, the consumption of market-based foods has been linked to increases in obesity, diabetes and cardiovascular disease in many Arctic communities (van Oostdam 2005). However, unlike traditional wildlife foods, market foods are monitored for contaminants through domestic policies and international trade laws; to this extent there is some assurance that these foods and food ingredients are safe for consumption. Kuhnlein and Chan (2000) further suggested the need for vigilance in monitoring the levels of persistent contaminants in the environment for their potential effects in the food chain and human health.

Overall, with climate change taking a toll in the Arctic-Barents region, it is important to have measures and strategies that will help mitigate these changes. Climate change outcomes leading to a shift away from traditional food is worrisome. It is better explained as cryospheric changes that will have consequences on local communities, national management and international regulations (Hovelsrud et al. 2012). Therefore, the importance is to maintain good governance across societal levels to facilitate adaptation and strengthen adaptive capacity. Additionally, social learning, responsive local institutions, livelihood flexibility, diversification and adaptive management have also emerged as critical elements.

It remains obvious that there are tremendous effects of climate change. These effects penetrate every aspect of life for people in the Barents region, and the impacts of the overall changes in the environment and animals clearly result in food insecurity within the region. Furthermore, climate change and human activities are detrimental to food security at both the individual and community levels. This is primarily seen through the cascading effects on health security, environmental security, community security, economic security and political security. Fortunately, humans are capable of planning ahead for unforeseen future contingencies that will affect food security.

6.6   The Relevance of the Region Regarding Global Food Insecurity in the Future

Historical records have shown that there are instances where adequate plans and the right conditions have prevented starvation and hunger. Such plans include preservation and storage techniques that provide humans with food when there are disasters. Svalbard (the Norwegian Arctic archipelago), within the Arctic-Barents region as the final frontier before the North Pole, is also acting as the last frontier in ensuring global food security for the future (Charles 2006; Siebert and Richardson 2011).

A recent initiative, "The Global Seed Vault," in Svalbard aims to respond to any sudden global calamity that can wipe out our common food supplies with seeds that form the foundation for nourishing our bodies. The seed vault is a storage capacity within the Barents region; the permafrost and extremely cold temperatures in Svalbard can help to preserve these seeds in the case of a failure in cold storage powered by electricity. The seed vault has the capacity to store 4.5 million varieties of crops, and each variety will contain on average 500 seeds, resulting in a maximum of 2.5 billion seeds that can be stored in the vault (Fowler 2008). Globally, more than 1700 gene banks hold collections of food crops for safekeeping, yet many of these are vulnerable, exposed not only to natural catastrophes and war but also to avoidable disasters, such as lack of funding or poor management. Currently, the Svalbard Vault holds more than 880,000 samples, originating from almost every country in the world. The collection ranges from unique varieties of main African and Asian food staples—such as maize, rice, wheat, cowpea and sorghum—to European and South American varieties of eggplant, lettuce, barley and potato (Fowler 2008). The seed vault represents the world's largest collection of crop diversity, and it safeguards as much of the world's unique crop genetic material as possible while avoiding unnecessary duplication. A temperature of −18 °C is required for optimal storage of the seeds, which are stored and sealed in custom-made three-ply foil packages. The low temperature and moisture levels inside the vault ensure low metabolic activity, keeping the seeds viable for long periods of time. The packages are sealed inside boxes and stored on shelves inside the seed vault. The permafrost at the vault was badly affected recently, as the end of 2016 saw average temperatures over 7 °C above normal, pushing the permafrost above the melting point, but precautionary measures were taken to make the vault waterproof (Carrington 2017).

Strengthening governance mechanisms and effective coordination across sectors can make a difference in solving global food insecurity challenges, such as natural resource degradation, globalization, high and volatile food prices, urbanization and climate change (FAO, IFAD and WFP 2014).

Chapter 7
The Governance of the Arctic-Barents Region and Food Security

7.1 Regional Governance Structure

The term governance refers to a system around which actors and institutions inter-act. This interaction offers a process wherein actors produce, for example, rules, regulations, policies, strategies and action plans. The process therefore ensures an overall direction in regard to how a particular issue is presented within a certain form of structure so that its effectiveness can be monitored or assessed. Generally, the governance of the Arctic and its Barents region is widely recognized as a system of fragmented international, regional and bilateral regulations. Such legal frame-work is complemented by non-binding policy tools, often referred to as soft law mechanisms; given that the region is apparently heavily institutionalized, a number of inter-governmental bodies are in place to address various concerns facing the region. The consensus among the Arctic states, in particular the five coastal states surrounding the Arctic Ocean, suggests no need for any comprehensive unique governance structure (Ilulissat Declaration May 28, 2008). The states suggest existing regulations, including the law of the sea along with the efforts of the Arctic Council, provide a sound basis for the governance of the region. As such, food security does not have any separate governance structure in the regional setting; rather, it is inte-grated within the existing governance framework as collections of several sector-specific regulation and policy tools directly or indirectly addressing issues having impacts on food security.

Therefore, food security is found to be one of the concerns directly or indirectly addressed in many policy instruments within the scope of these legal tools and insti-tutional bodies. All four countries within the Arctic-Barents region are eager to exploit the natural resources located within the region, wherein food contributes as an important element. The regulations, strategies and policy instruments developed at both national and institutional levels emphasize that the exploitation of resources is to take place in a sustainable manner (AMAP 2017a, b). Both regulations and policy instruments play an important role in the sustainable exploitation and management of food resources with a view to ensuring greater food security in the regional setting. In addition, the international human rights framework as a norm of governance applicable to the region in connection to food security also provides mechanisms for the governance of food security, in particular as it relates to both the right to food as a general human rights norm and the right to food as an element of culture, in particular for indigenous peoples. This chapter, therefore, offers a picture of the governance framework, highlighting various regulatory and policy processes as they exist in connection to food security. In addition to the existing bodies of law applicable to food security, we highlight four institutions that play a significant role in the Arctic-Barents region in one way or another to promote food security: the Arctic Council, Northern Dimension, the Barents Euro-Arctic Council and the EU. These four institutions offer significant policy tools contributory to food security in the Arctic-Barents region.

7.2　Climate Change Strategies as They Relate to Food Security

Climate change and its effect on food security are evident in many ways. There are a number of studies suggesting the effect of climate change on food security (e.g., Wheeler and von Braun 2013). Changes in temperatures and precipitation have the potential to alter the distribution of agro ecological zones. Agricultural losses can result from climate variability and the increased frequency of extreme events such as droughts and floods or changes in precipitation and temperature (Zewdie 2014). Climate change implications are also linked to CO_2 emissions, biodiversity loss, water scarcity and higher economic costs, eventually leading to adverse effects on health-enhancing, sustainable diets (Lang 2017). While the potential impacts of climate change on the global food system are relatively less clear, studies suggest that the stability of the whole food system is at risk due to variability in short-term food supply (Wheeler and von Braun 2013). It is therefore important to strengthen the strategies to improve both mitigation and adaptation techniques to create a "climate-smart food system," which is more resilient to climate change influences on food security (Wheeler and von Braun 2013).

　Climate change strategies in relation to food security are not separately addressed in the regional context. Rather, several of the policy instruments and strategies offer overarching features as they relate to food security. As analyzed elsewhere in this book, climate change and its consequences provide significant challenges in the region, which affect, among others, food security. As a result, climate change strategies at various levels contribute to the promotion of food security and the safety of food consumption in the region. In the following sections, therefore, we highlight some of the climate change strategies applicable to the region and in relation to food security.

7.2.1   EU Climate Change Strategy

The Arctic-Barents region is closely connected to the EU; two of the countries—Finland and Sweden—are EU members, and Norway is closely linked to the EU via the European Economic Area (EEA) agreement. In addition to countries such as Finland, Norway and Sweden, both Russia and the European Commission are also the members in the Barents Euro-Arctic Council. As a result, the EU has great potential to influence the region through, for example, its climate change strategy. The EU climate change legislation and framework are of importance when climate governance applicable to the region is at stake. The EU framework has set a timeframe in the "2020 energy and climate package" and the target of "20-20-20," which refers to increasing the share of renewable sources of energy by 20%, improving energy efficiency by 20% and reducing GHG emissions by 20% (EU 2016). The EU-level regulatory mechanisms are in place to assist the member countries in meeting the goals (Braun 2011).

In addition, the Common Agricultural Policy 2014–2020 has measures to increase the carbon sink by encouraging more grassland and the protection of forest cover as well as the mitigation of challenges to promoting soil quality (GRICCE 2016). The EU Renewable Energy Directive is worth mentioning in this context, as it contributes to meeting the 2020 goal by increasing energy use from renewable sources, with prescriptions for the member states to ensure that at least 10% of their transport fuels come from renewable energy sources by 2020 (EC-RED 2016). The overall goal of the Directive is to limit greenhouse gas emissions and remove interference issues, such as those in global food production (GRICCE 2016). The European Commission is expected to consider the proposal of a legally binding instrument in 2017 due to the fact that individual member states' actions are deemed insufficient to meet the overall 2020 goal. Such regulation is also expected to better coordinate the climate change national adaption policies of the member nations (GRICCE 2016). This could see further policies being handed down from the EU-level form of governance, hopefully enhancing cooperation on additional climate change strategies and policies that further protect and promote food security.

Looking into the future, policy makers face a new and challenging set of issues: how to develop strategies for fighting against new environmental problems, how to develop better strategies for solving the old ones and how to do both in ways that are more efficient and less taxing and that engender less political opposition (Kettle 2002). The strategies, legislation and actions taken by countries in the Barents region have been found significant for confronting the adverse effect of climate change, which then, among others, contributes to the promotion of food security.

7.2.2   Regional Climate Strategies

The Arctic-Barents region is Europe's largest region for interregional cooperation, and this cooperation is organized on two levels, national and regional (Himanen et al. 2012). In December 2013, the Barents Environmental Ministers' Meeting adopted the Action Plan on Climate Change for the Barents region. This Action Plan incorporated regional climate strategy plans for the Barents region. The regional climate change strategies are important tools for addressing the mitigation of and adaptation to climate change, as they can be used to consolidate the efforts of different stakeholders in the public and private sectors (Himanen et al. 2012). They are also useful toward achieving national and international goals for the reduction of greenhouse gas emissions. Most of the Barents region is either covered by or in the process of being covered by a climate action strategy. The importance of a climate plan lies in the further mitigation of climate change having impacts on food security in the region. Furthermore, regional action plans addressing climate change can be used as a template for further action plans in regard to food security.

Finland, Sweden, Norway and Russia have all implemented goals for their different counties within the region, including a focus on a number of sectors within the region to reduce GHG emissions. The majority of counties within the region already have a plan in place; while not legally binding, these plans do serve as a step toward a solidified strategy in the future. Although counties within the Barents region have their own plans, many of the counties in Russia are still far behind. The Russian part of the Barents faces difficulties with regard to funding, since the Barents Cooperation does not have its own financing instruments. Funds for the implementation of investments in energy efficiency, black carbon mitigation and other infrastructure-related projects are available from international financial institutions (Sorvali 2015). Tennberg notes that Sweden and Norway take a local approach to developing their climate plans with a focus on municipalities, whereas Russia and Finland focus more on the regional approach. Tennberg also points out that during meetings with officials in developing a regional climate plan, issues of food security have not been brought forward or discussed (Interview with Monica Tennberg 2016).

Plummer and Baird noted that both opportunities and challenges will emerge from climate change in the Barents region. Responses that realize the former and minimize the latter are highly desirable, and it is anticipated that approaches stressing cooperation on, and adaptation to, climate change will be the key strategy (Plummer and Baird 2013). Both Plummer and Baird suggest co-management as a possible governance approach for climate change adaptation in the Barents region, but this could be challenged when confronted by multi-level governance. They explained that while adaptive co-management typically concentrates on the local or regional level and emphasizes linkages to actors both horizontally and vertically, international forces may overcome it (Plummer and Baird 2013). They also use the example of policies set out by the EU regarding slaughter facilities in the Barents region. This policy, implemented at a higher level, extends to the lower levels, forcing reindeer herders to travel further or ship their reindeer (Plummer and Baird 2013). As discussed with Juha Joona, these policies are designed for herders wanting to sell their meat to the general public in stores and markets, which now requires that they have the meat butchered in one of these designated facilities (Joona 2016a, b). Plummer and Baird gave another example, discussing the isolated Pomors fishing villages on the western coast of the White Sea, where local sustainable fishing practices have been jeopardized by federal laws aimed at promoting industrial fishing and increasing tourism in the region (Nystén-Haarala and Kulysasova 2012).

A climate change strategy for the region could be a success in the Arctic-Barents region. It seems that when a climate strategy exists, the incorporation of climate change-induced issues into other policy documents and processes follows (Sorvali 2015). One thing that is not directly mentioned in such strategies, but which is closely associated, is food security. Food security should be tied in with these climate change strategies and identified to reduce harm to individuals and animals. Furthermore, such progress in a climate change strategy might also advance toward a regional strategy on food security.

7.2.3   National Strategies

Countries in the Barents region also have their own national action plans and strategies toward climate change. Each country has a series of commitments that they have followed. For the EU members in the region, sometimes the action plans have been handed down from the EU. For others, the plans have been based on international commitments. Although legislative documents on climate change might not be directly related to food security in the Barents region, they do hold food security implications due to the nature and impact of climate change on food security in the region.

Russia as a country has committed to reducing its emissions of net greenhouse gases (GHG) by 25–30% below the 1990 level by 2030 (Russian Federation 2016). The country has also implemented a number of legislative documents that could be seen as actions toward climate change mitigation and adaption. The Greenhouse Gas Emission Reduction of 2013 establishes a target for emissions reduction, stating that by 2020 emissions cannot exceed 75% of the total emissions from 1990 (GRICCE 2013a, b). The Climate Doctrine of the Russian Federation (2009) sets strategic guidelines and serves as a foundation for a future climate policy. Overall, these actions can be seen as small in comparison to other countries in the Barents region, but they have come a long way since the collapse of the Soviet Union.

Norway has a number of initiatives on climate change. Even though Norway is not an EU member, it is, however, still bound to a large degree by EU legislation through the Agreement on the European Economic Area (EEA). Norway has a Climate Settlement (Innst. 390 S 2011–2012), which is not really considered legislation, but it does guide and set the framework for discussions around climate change (GRICCE 2013a, b). In addition, Norway has set non-binding standards by which it expects to be carbon neutral by 2050 or earlier. However, it aims to reduce GHG emissions by 30–40% by 2020 from the 1990 emissions baseline (NEA 2015). In 2013, Norway adopted the "White Paper on Climate Change Adaptation" (Nyborg 2013), which provides concrete guidelines concerning emissions reduction. The Government of Norway plans to use the white paper to conduct regular assessments of vulnerability and adaptation needs in Norway.

Finland took a stance in 2015 to commit to an 80% emissions reduction by 2050 using 1990 as a baseline through its Finnish Climate Change Act (2015). By virtue of this regulation, Finland aims at creating a framework for bottom-up, long-term, consistent and cost-effective climate policy planning and the implementation of a low-carbon society (VTT 2012). The country has particularly addressed the sectors of transport, agriculture and housing as well as those under the EU emissions trading system (ETS) directive (electricity production, energy-intensive industry, a large share of district heat production and aviation). Furthermore, the Flood Risk Management Act was adopted in 2010, the aim of which is to reduce flood risks, prevent and mitigate the adverse consequences caused by floods and promote preparedness for floods (FEA 2015). While both Finland and Sweden rely heavily on EU regulations and directives, their commitment to the Kyoto Protocol and the subsequent Paris Agreement and international emission reduction targets is worth mentioning (GRICCE 2013a, b). In Sweden, as in Norway, smaller municipalities also play a role toward meeting international climate mitigation commitments, with policy measures on areas such as energy management, land use, waste and transportation (Nachmany 2015a, b). In 1999, Sweden implemented the Environmental Code DS2000: 61, which contains 15 separate environmental acts into this one code concerning the management of land and water areas, environmental quality standards, the protection of nature and polluted areas, to name a few (SEPA 2016).

7.3   Legal Tools Applicable to the Arctic-Barents Region as They Relate to Food Security

7.3.1   UNFCCC, Kyoto Protocol and Paris Agreement

The United Nations Framework Convention on Climate Change (UNFCCC) entered into force in 1994 with an aim to address measures to mitigate climate change. The UNFCCC is primarily focused on the "stabilization of greenhouse gas (GHGs) concentrates in the atmosphere at a level that would prevent dangerous anthropogenic interference with the climate system" (UNFCCC 1992). As discussed, mitigating climate change does have implications for food security. One of the aspects the UNFCCC presents is about ensuring supportive climatic conditions for food production and allowing economic development to proceed in a sustainable manner (UNFCCC 1992). Food is a global issue, but the consequences of GHG emissions are causing it to become a reason for concern for many regions across the globe.

While the UNFCCC provides a general framework concerning the mitigation of climate change, the Kyoto Protocol, adopted in 1997 within the framework, provides for concrete commitments by states for internationally binding emission reduction targets (Depledge 2000). As a framework, the UNFCCC remains a widely recognized and influential instrument due to its high ratification rate among the states. It should be noted that all eight circumpolar Arctic states are parties to the Convention (Nowlan 2001). However, the United States has never ratified the Kyoto Protocol, and Canada withdrew itself from the Protocol in 2011 (EC Canada 2012). Given that the targets within the Kyoto Protocol have not been effectively complied with by many of the states and that effective implementation of the emissions reduction requires stronger measures, the Conference of Parties (COP) 21, in 2015, negotiated a new treaty—the so-called Paris Agreement.

The Agreement came into force in November 2016, in less than a year's time. To date, 151 states have ratified the Agreement. It has a particular focus toward the mitigation and adaptation of climate change with renewed targets. The primary aim of the Agreement is to keep the global temperature rise well below 2 °C during this century and to drive efforts to keep it to 1.5 °C above pre-industrial levels. In the COP 21, there is explicit reference made within the preamble of the Agreement to food security and production, which acknowledges "the fundamental priority of safeguarding food security and ending hunger, and the particular vulnerabilities of food production systems to the adverse impacts of climate change" (OECD 2016a, b). In addition, the preamble includes references to human rights, development, gender, ecosystems and biodiversity, all of which are of key importance to agriculture. Article 2 of the Agreement underscores the importance of food production, clearly stating that the need for strengthening the global response to climate change in a manner that does not threaten food production. The Agreement also acknowledges the importance of fostering climate resilience through low greenhouse gas emissions, which implicates the need to re-fashion food systems in a sustainable manner.

Ultimately, the Agreement is expected to better safeguard climatic conditions so that the promotion of food security is better maintained. Among the Arctic states, all but Russia have ratified the Convention. Thus, the Agreement only covers the Barents region partly—the vast area of the Russian Barents is left outside of the scope of the Agreement. Food security does present an explicit concern under the climate change regime. Successful adaptation and mitigation responses can only be achieved within ecologic, economic and social sustainability. The regulations pertaining to climate change are about the promotion of sustainable ecological processes, which eventually provide life support systems for animals, plants and human beings. Thus, addressing the effects of climate change and related environmental and ecological problems both directly and indirectly present concerns of food security.

7.3.2 United Nations Convention on the Law of the Sea

The United Nation Convention on the Law of the Sea (UNCLOS) provides an overarching legal framework for the governance of the world's oceans and seas. The UNCLOS is thought to be one of the most relevant and comprehensive documents that applies to the Arctic, in particular in its marine area. It is considered to be the most referred to legally binding framework for the Arctic region given that the Arctic Ocean is surrounded by five coastal states and that these states have clear jurisdictions within both the water column and the continental shelf of the Arctic Ocean (Fallon 2012). The UNCLOS, even though adopted in 1982, came into force in 1994. All of the Arctic states except the U.S. are parties to the Convention. Even though the U.S. is not a party to the UNCLOS, it is considered to be bound by the customary law of the sea, which is embodied within the framework of the UNCLOS. Moreover, the U.S. is a member country of the Arctic Five—the so-called club of the five Arctic coastal states. In 2008 these five coastal states adopted the frequently referred to Ilulissat Declaration, by which these states recognized the relevance of the law of the sea and the UNCLOS as the central legal instrument for Arctic governance (Dodds 2013). As for the Barents region, the whole of the region's marine areas are covered by the UNCLOS, as all the countries in the region have endorsed the instrument as binding.

The usage of the marine areas is far reaching, extending beyond conventional purposes, and can affect human lives. Most coastal communities in the Barents region rely on marine-based resources, in particular for ensuring the supply of food, such as fisheries. Many coastal communities, especially the indigenous communities, also rely on sea ice to use as ground for hunting purposes. According to Nuttall et al. (2005), while the indigenous communities across the Arctic-Barents continue to depend on the harvesting and use of living terrestrial resources, they also rely heavily on marine and freshwater resources. Similarly, such dependence is also found in other communities living near the coastal areas. Apart from the exploitation of marine life resources, the waterways of the Arctic-Barents region are used for transportation, especially during the summer months when the ice melts and the marine areas are open for a longer period. Such usage of marine areas is argued to be detrimental to the marine environment, precisely in the context of this specific region given that the region's ecosystem is highly sensitive. The exercise of resource usage both in surface and subsurface water as well as in the seabed of the continental shelf may have various consequences, the most alarming of which is the risk concerning the protection and preservation of the marine environment. Living resources can be overexploited, resulting in the destruction of the ecological balance. The extraction of non-living resources—that is, hydrocarbons exploitation—may cause extensive pollution to the marine environment if, for example, an accident occurs in the process of extraction or transportation. In addition, other human activities, such as shipping, may also contribute to the pollution affecting marine living resources. As transportation through marine areas or other commercial usage of the marine areas has potentials for bringing pollution, food derived from the marine ecosystem is found to be contaminated (AMSA 2016). The species most commonly harvested from the marine environment in the Barents region include those such as fish, whales, seabirds and seals. These species, having been exposed to contamination, eventually affect food security, resulting in an effect on human health. Further, a broader range of anthropogenic threats affects marine biodiversity, causing the depopulation and migration of species.

The UNCLOS is not an environmental treaty by any means, but it does make significant references to the environment, in particular concerning the protection of the marine environment. A full chapter of the UNCLOS is dedicated to marine environmental governance (Part XII). This part lays out both the general and specific obligations for states in terms of protecting and preserving the marine environment as well as controlling marine environmental pollution when extracting living and non-living resources from the areas that belong to national jurisdiction. States are also obliged to pay due regard to environmental protection in all maritime areas, including the high seas. States' duty also lies in the prevention, reduction and control of pollution arising from land-based sources (Article 207, UNCLOS). By virtue of the provision embodied in Part XII of the UNCLOS, states also incur legal responsibility to never cause marine pollution detrimental for other states (Hermeling et al. 2015; Porta et al. 2017). Given its comprehensive set of rules, the UNCLOS is referred to as an international "constitution" on the protection and preservation of the marine environment (Roberts 2007).

As it relates to the Barents-Arctic, one of the UNCLOS provisions is particularly relevant for the region, namely Article 234. This Article provides the coastal states with the right to adopt and enforce non-discriminatory laws for the prevention and control of marine pollution from vessels in ice-covered areas. Thus, given that the status of the marine area in the region is treated as ice-covered in a legal sense, the coastal states have the duty to prevent marine pollution from causing major harm to, or irreversible disturbance of, the ecological balance (UNCLOS 1982). While this Article has clear relevance as it relates to food security, the UNCLOS, however, does not provide any reference to indigenous and coastal communities nor to perspectives in regard to the human rights of these groups of peoples. Failure to address the concerns of traditional livelihood practices in relation to, for example, the food consumption practices of these communities, especially in the Arctic-Barents region, creates a gap in the food governance framework in the course of the changing environment (Fallon 2012).

7.3.3 Convention on Biological Diversity

The Convention on Biological Diversity (CBD) is an overarching treaty aimed at the preservation and conservation of global biodiversity. This is the first treaty of its kind concerning the protection and conservation of ecosystems on a global scale (Nowlan 2001). Biodiversity in the Arctic-Barents is important and is often compared to the Amazon as one of the most ecologically diverse places on Earth (Picq 2012). The Convention highlights two key themes: the sustainable use of biological resources and the equitable sharing of benefits derived from the use of biological resources (CBD 1992). In addition, the Convention also addresses the protection of marine biodiversity. In its preamble, the CBD directs the parties to adopt a precautionary approach (Mace and Gabriel 1999), which is then reiterated by COP Decision II/10 concerning the marine environment (CBD 1995). Although the CBD does not provide additional guidance on the nature and scope of this mandate, the Arctic coastal states that are parties to the Convention, like all other state parties, are necessarily guided by the principle of the precautionary approach in marine resources governance (UNFSA 1995). In addition, CBD also offers the creation of Marine Protected Areas (MPAs), through which a wide range of human activities can be regulated to protect marine species, including their conservation (Roberts 2007). Food is a good example of a biological resource in the Barents region. People inhabiting the region—both indigenous and non-indigenous—rely on the region's natural resources, which they use in a sustainable manner. Protecting such biological resources will lead to further securing traditional and local food systems. Across the Arctic and among Arctic-Barents peoples, it is believed that the harvesting of wild species is the single most common feature of natural resource use (Nowlan 2001). Therefore, for the peoples of the Barents region, food is a precious resource that must be preserved and protected.

While the CBD as a whole is significant for protecting and preserving the region's rich biodiversity, Article 8 is of particular importance where food security is concerned. Article 8 touches upon a number of issues, such as the prevention, control, introduction and eradication of alien species having negative impacts on biodiversity; the maintenance and preservation of knowledge, innovations and practices embodying traditional lifestyles relevant for the conservation and sustainable use of biological diversity; the promotion and wider application of such knowledge, innovations and practices with the approval and involvement of their holders; and encouraging the equitable sharing of the benefits arising from the utilization of such knowledge, innovations and practices. The Convention also highlighted the need for the adoption of necessary legislation and/or other regulatory provisions for the protection of threatened species and populations.

Generally, food is voiced with awareness, where the conservation and sustainable use of biological diversity is of critical importance for meeting the food, health and other needs of the growing world population and with the utmost significance for survival (Dulloo 2013). People in the Barents region have practiced the sustainable use of biological diversity over many centuries, primarily for consumption and survival as well as for traditional and cultural needs. According to Nowlan, preserving cultural values and the traditional lifestyle requires safeguarding the natural environment, and, by protecting the natural environment, Arctic-Barents communities are able to receive both conservation and economic benefits from the use of the region's natural resources, including local and traditionally used food (Nowlan 2001). The CBD also highlights the adoption of national legislation and regulatory provisions for the strengthened protection of biological resources. With the exception of the United States, all the Arctic states, including the whole of the Barents region, is covered by the Convention.

7.3.4   The OSPAR Convention

The Convention for the Protection of the Marine Environment of the North-East Atlantic (OSPAR) is a regional legal instrument that generally covers the North-East Atlantic. However, the Convention applies to some parts of the Arctic-Barents waters, too. For example, its "Region 1" covers both the Norwegian and Russian parts of the Barents Sea. Even though Russia is not yet a party to this Convention, it provides a standard for the eco-system based management of marine areas (Hansen et al. 2016). The objective of the Convention is to conserve marine ecosystems and safeguard human health by preventing and eliminating pollution. It also aims at protecting the marine environment from the adverse effects of human activities while it offers measures for sustainable use of the seas. Given that the marine diversity in this area is an important source of food for nearby communities, the Convention offers a marine environmental governance supportive of ensuring marine food security and setting a standard for other marine areas in the Arctic-Barents region.

7.3.5 Minamata Convention on Mercury

The Minamata Convention on Mercury—referred to as the Mercury Convention—was adopted in the year 2013 and became effective in August 2017. This is the most recent legal instrument for the protection of human health and the environment from anthropogenic emissions and releases of mercury and mercury compounds (Minamata Convention 2017).

Mercury (Hg) is a harmful substance that has overwhelmed people and food sources in the Arctic for years. This is because the pristine Arctic region is today regarded as a special case for mercury—about 200 tons of mercury is deposited in the Arctic annually, generally far from where it originated (Kirby et al. 2013). People living in the Arctic-Barents region can be distressed from the pollution resulting from mercury. This region's indigenous peoples especially in coastal areas are far more susceptible to these toxins as a result of their traditional and local food sources. According to the 2011 Arctic Monitoring and Assessment Programme (AMAP) report, mercury levels are continuing to rise in some Arctic species used for food sources, despite reductions over the past 30 years in emissions from human activities in some parts of the world (AMAP 2011a, b). The report also documented a ten-fold increase over the last 150 years of the mercury level in species, such as belugas, ringed seals, polar bears and birds of prey (AMAP 2011a, b). Over 90% of the mercury in these animals, and possibly some Arctic human populations, is believed to have originated from human sources (AMAP 2011a, b).

The Mercury Convention applies to most parts of the Arctic, as most of the Arctic states have ratified the Convention. However, among the Arctic states, Iceland and the Russian Federation are not yet parties to this Convention, thus, the Barents region is not fully covered thereby. The Convention nevertheless sets a legal standard in relation to food security governance in the region.

7.3.6 Shipping-Related Legal Instruments Applicable to Food Security

One of the major changes the Arctic-Barents region is expected to face in the near future is an increase in shipping activities across the emerging Arctic sea routes. The Arctic sea routes are argued to be shorter, energy and time savers and relatively safer compared with the traditional sea routes. According to statistics, the Northern Sea Route along the Russian coast has seen rapid growth in recent years (Liu and Hossain 2017). Although the harsh climatic conditions are known to delay regular shipping operations, new technological developments, such as shipping with ice breaking facilities, suggest a gradual progression. Moreover, many areas in the Barents region, especially along the Norwegian coast and Murmansk in Russia, are already ice-free year round due to the Atlantic Gulf Stream (Rottem and Moe 2007). Vessel-sourced pollution does have the potential to contaminate food sources in the marine areas of the Arctic-Barents region. As a result, the existing international and regional legal instruments addressing shipping are relevant in the regional context to offer governance of food security. The following sub-sections present some of these instruments relevant for the Arctic-Barents region.

7.3.6.1   Convention on the Prevention of Marine Pollution by Dumping (London Convention)

In 1972, the Convention on the Prevention of Marine Pollution by Dumping of Waste and Other Matter, known as the London Convention, was adopted (London Convention 1972). Article I of the London Convention mandates that states take all practicable steps to prevent the pollution of the sea by the dumping of waste and other matter that is liable to create hazards to human health, to harm living resources and marine life, to damage amenities or to interfere with other legitimate uses of the sea (London Convention 1972). This document is of particular use to the Arctic environment, since the region has been used as a dumping ground for hazardous wastes (Rothwell 2000). The dumping of hazardous wastes in the Arctic-Barents region was in fact common during the Cold War in the Soviet Union era. The dumping ground location was around the Barents Sea and Novaya Zemlya in Northwestern Russia (Pursiainen 2001). The International Atomic Energy Agency (IAEA) produced a report focusing on radioactive dumping in the Arctic seas, and the report made it clear that the gradual deterioration of the waste packages and containers could lead to adverse impacts in the future (IAEA 2003). The report noted that the dumping as such could result in the contamination of the marine food chain, with the possibility of additional radiation exposure to humans through the consumption of fish and other marine foodstuffs as a consequence (IAEA 2003). These findings remain unclear, because the effects are still widely unknown, but studies are progressing to promote the understanding of the impacts.

The Convention has been ratified by all the Arctic states and thus applies to its Barents region. In 1996, a separate protocol—the London Protocol—was agreed upon with a view to modernize the Convention and, ultimately, to replace it. Under the Protocol all dumping is prohibited, except for possibly acceptable wastes on the so-called "reverse list" (London Protocol 1996). The Protocol came into force on March 24, 2006, and as of December 2016 there are 48 parties to the Protocol. Among the Arctic countries, except Finland and Russia, all states are parties to the Protocol. Thus, the prohibition of all dumping, with the exception of those indicated n the Protocol, does not apply to the Barents region as a whole, resulting in short-comings for food security governance in the regional setting.

7.3.6.2 International Convention for the Prevention of Pollution from Ships

Similar to the London Convention, the International Convention for the Prevention of Pollution from Ships (MARPOL) of 1973, supplemented by the 1978 Protocol and binding since 1983 (MARPOL 2016), addresses minimizing the pollution of the oceans and seas from dumping and oil and air pollution. The MARPOL is the most gold-standard international convention that oversees the prevention of pollution to the marine environment by ships due to operational or accidental causes (MARPOL 2016). The protection of oceans and seas from vessel-sourced pollution benefits the living marine resources by preventing their contamination. The protection as such offers benefits to coastal communities relying on food sources from the marine environment and thus contributes to food security governance. The Convention binds all the Arctic states and covers the whole of the Barents region.

7.3.6.3 Ballast Water Convention

The International Convention for the Control and Management of Ships' Ballast Water and Sediments (BWM) was adopted in 2004 and came into force in September 2017. The aim of the Convention is to prevent the spread of harmful aquatic organisms from one region to another. The Convention thus promotes standards and procedures for the management and control of ships' ballast water and sediments (BWM 2004). The Convention compels signatory nations to ensure by 2016 that all ballast water in both old and new ships is treated before being discharged. Furthermore, it requires that by 2012 all new ships treat their ballast water. Prior to 2012, all vessels were required to discharge their ballast water in the open sea (Hansen et al. 2016). Ballast water is a deep concern for invasive species entering the Arctic-Barents region, as they present a major threat to marine ecosystems. According to the Arctic Marine Shipping Assessment Report of 2009, the expansion of maritime traffic in the region increases the possibility of introducing alien species and pathogens from ships' ballast water discharge and hull fouling (AMSA 2009). The introduction of alien species poses particular threats not only to the marine environment as a whole but also to the marine food chain on which Arctic-Barents communities rely. The Convention thus set standards that food security governance benefits from, given that the ecosystem in the Arctic-Barents region is highly sensitive to any detrimental effect from the introduction of alien organisms.

7.3.6.4 Polar Code

The Polar Code is the most relevant shipping-related legally binding document that applies to the Arctic. The Polar Code amended Annexes I, II, IV and V of MARPOL and introduced the new Chapter XIV within the framework of its International Convention for the Safety of Life at Sea (SOLAS). The measures, effective as of 2017, focus on safe vessel operation and protection of the marine environment in polar waters by addressing the risks present in polar waters and not adequately mitigated by other instruments adopted within the framework of the International Maritime Organization (IMO) (Polar Code 2014). The instrument safeguards the Arctic environment in a number of ways—by putting forth protective measures against oil pollution, the introduction of invasive species, sewage, garbage and chemicals. The Polar Code is important for ensuring sustainable shipping and to prevent detrimental impacts on the environment linking to, for example, food and marine living resources. As mentioned earlier, increased marine traffic in the Arctic-Barents creates risks for the region's marine environment. The Polar Code ensures safeguards against specific problems prevailing in, for example, Arctic waters, and the Code essentially contributes to the promotion of food security in regional settings.

7.3.7 The Convention on Long-Range Trans-Boundary Air Pollution

The Convention on Long-Range Trans-Boundary Air Pollution (LRTAP) was adopted in 1979. The Convention is the first internationally legally binding agreement outlining the principles for regional cooperation on trans-boundary air pollution (UNECE 2016). The Convention has been supplemented by three additional protocols. The first is the 1994 Oslo Protocol, which came into force in 1998, aiming to reduce sulfur emissions (Oslo Protocol 1994). Second, the 1998 Aarhus Protocol on Heavy Metals, entered into force in 2003, required the reduction of three harmful metals: cadmium, lead and mercury emissions (UNECE 1998). Parties to the Protocol are obliged to reduce emissions from industrial sources, combustion processes and waste incineration while using the best available techniques (BATs) for stationary sources, such as the use of special filters, scrubbers or mercury-free processes. Finally, complementary to the Protocol on Heavy Metals, the 1998 Aarhus Protocol on Persistent Organic Pollutants (POPs) entered into force in 2003. The Protocol was further amended in 2009. However, the amended version has not yet entered into force. The objectives of this Protocol are to control, reduce or eliminate discharge, emissions and losses of persistent organic pollutants. The Protocol stipulates the elimination or reduction of the production and use of 13 substances regarded as persistent organic pollutants (UNECE 1998). In addition, parties are required to take effective measures to reduce or stabilize the total annual emissions of certain substances.

Long-range air pollution has been at the forefront of issues in the Barents region, as such pollution has had a tremendous impact on the environment and food sources both on land and in water. Given the sensitive environmental conditions prevailing in the region, emissions of sulfur and other pollutants, such as heavy metals as well as POPs, contribute to adverse effects in the region. For example, POPs are transported far from their site of origin, where the atmosphere is a dominant medium for such transportation. Approximately half the substances targeted in the POPs Protocol are not subject to immediate elimination. Unless the emissions are controlled, the pollutants addressed in the Convention as well as in its subsequent protocols present threats to not only the environment in general but also to humans, animals and other species. Since in the region people and communities rely on the supply of the traditional food chain, the contamination from these pollutants does indeed have the potential to affect food security. Therefore, both the Convention and the subsequent protocols provide mechanisms to safeguard food security. However, not all the Arctic states have ratified these protocols, yet all are parties to the LRTAP. Concerning the ratification of the protocols, VanderZwaag highlighted that despite the slow process, the mechanism has opened up the doors for further legally binding documents concerning pollution occurring from these sources (VanderZwaag 1998).

7.3.8 Stockholm Convention on Persistent Organic Pollutants

The Stockholm Convention on Persistent Organic Pollutants (POPs) was adopted in 2001 and entered into force in 2004. The Convention stressed global action to reduce the harmful effects of POPs—the chemical substances that persist in the environment and bio-accumulate through the food web, which eventually result in threats to human health and the environment. The Convention is particularly relevant in the context of the Arctic-Barents region. For example, the preamble to the Convention clearly acknowledges the vulnerability of Arctic ecosystems, and especially that of indigenous communities, who are at particular risk because of the bio magnification of POPs and, in particular, the contamination of traditional foods (Stockholm Convention 2004). POPs are well documented in the Barents region and a main source of food pollution in the region. Therefore, the Convention's banning of such pollutants greatly complement the region's food security governance. As of today, except for the U.S., all Arctic states have ratified the Convention, hence making the entire Barents region bound by treaty provisions.

7.4   General Overview of Institutions Governing the Region in Relation to Food Security

7.4.1   Arctic Council

The Arctic Council (AC) is a high-level inter-governmental organization of the eight Arctic states governing the Arctic and its Barents region (Ottawa Declaration 1996). The structure of the AC includes not only the Arctic states but also the region's indigenous peoples represented by their respective organization, called "permanent participants" (Ottawa Declaration 1996). These permanent participants are fully consulted in all deliberations and activities of the Arctic Council (AC 2016). In addition, the AC structure includes observers consisting of non-Arctic states and other actors, such as international non-governmental organizations. The functions of the AC are generally conducted through six working groups, namely the Arctic Contaminants Action Program (ACAP), the Arctic Monitoring and Assessment Programme (AMAP), Conservation of Arctic-Barents Flora and Fauna (CAFF), Emergency Prevention, Preparedness and Response (EPPR), Protection of the Arctic Marine Environment (PAME) and the Sustainable Development Working Group (SDWG). The resolutions adopted within the framework of the AC are primarily treated as soft-law documents given that the institution does not intend to produce law. However, the documents produced within the framework of the AC do have an influential effect given that they are grounded in science-based knowledge (IASC 2008). These documents complement the issues embodied in the legally binding documents discussed in the previous section.

In relation to food security, the AC, through its SDWG and AMAP, produced a report in 2013 on food and water security indicators in the Arctic as they relate to health (Nilsson and Evengård 2013). The report explores a number of indicators having potential to adversely affect human health all across the Arctic, including its Barents region. While this report directly addresses food and water, there are other similar documents produced within the framework of the AC that are directly or indirectly relevant in regard to food security. For example, the Arctic Marine Shipping Assessment report (2009) was produced within the auspices of the PAME working group, which recognizes indigenous peoples and their reliance on the ocean and marine environment for subsistence. The report highlighted the various consequences of shipping usages in the Arctic that could impact the marine environment and its species, such as marine mammals used as traditional sources of foods, as in fisheries, for indigenous and coastal communities (AC 2009).

The Council, however, has not yet focused on the conservation and management of targeted species nor does it have any working group with a mandate to deal with these issues. For example, to deal with fishery issues, the CAFF and SDWG produced the Arctic Biodiversity Assessment report (ABA report 2013) and the Best Practices in Ecosystems-Based Ocean Management report (Hoel 2009), respectively, which are useful for the conservation and management of fisheries. Moreover, the combined efforts of AC's working groups—PAME, EPPR, AMAP and CAFF—have been involved in developing the regulatory framework to minimize the pollution occurring from offshore extractions. In 2009, the AC endorsed the Arctic Offshore Oil and Gas Guidelines (AOOGG 2009), recognizing a uniform understanding of the minimum actions needed to protect the Arctic marine environment from unwanted environmental effects caused by offshore oil and gas activities. The guidelines highlight the requirement of having an environmental impact assessment (EIA) procedure and plans for emergencies and responses.

7.4.2   Northern Dimension

The Northern Dimension (ND)—a body founded in 1999 that was renewed in 2006—is a joint policy between the EU, Russia, Norway and Iceland (ND 2016). The policy of the ND aims at supporting stability, wellbeing and sustainable development in the region by means of practical cooperation (ND 2016). It covers a wide range of sectors, such as the environment, nuclear safety, health, energy, transportation, logistics, the promotion of trade and investment and education and culture, to name a few (ND 2016). The policy of the ND is directed toward providing a common framework that promotes dialogue and concrete cooperation among the actors. Strengthening regional stability, providing wellbeing and intensified economic cooperation and integration and ensuring competitiveness and sustainable development in Northern Europe are the primary goals of the policy (ND 2016). The ND policy does not deal directly with food security issues; however, its works focus on environmental cooperation, which is relevant for furthering food security and safety in the region. The cooperation within the framework of the ND can be used as a platform for the promotion of dialogue and cooperation on issues such as climate change and its effect on food security in regional settings.

7.4.3   Barents Euro-Arctic Council

The Barents Euro-Arctic Council (BEAC) is an inter-governmental forum consisting of member countries including Denmark, Finland, Iceland, Norway, Russia, Sweden and the European Commission (Arctic Portal 2017). In addition, BEAC has nine observers: Canada, the U.S., the U.K., the Netherlands, Germany, France, Italy, Japan and Poland (Barents Cooperation 2017). The cooperation within the framework of BEAC was established in 1993 and was conducted at the inter-governmental level as well as at the inter-regional level. While inter-governmental cooperation is performed by governmental representatives, inter-regional cooperation is performed through the county representatives of 14 counties within the whole Barents region (BEAC 2016a, b, c). In addition, the indigenous peoples of the Barents region, such as the Saami, Nenets and Veps, are heavily engaged within the framework of BEAC. Like the Arctic Council, BEAC's functioning is conducted through a number of working groups at both the inter-governmental as well as inter-regional levels. Moreover, there are six working groups under a separate "joint working group" category in which both government and regional representatives participate. It is also important to note that the Barents Cooperation also includes a fourth category of working group—the indigenous peoples' working group—created under the auspices of its regional council to promote the rights of the indigenous peoples of the region. The Saami, the Nenets and the Vepsian people comprise this working group. The objectives of BEAC as a whole are to cooperate in a number of areas, including protecting the region's environment from degradation; promoting emergency and rescue services; promoting the wellbeing of youth groups; protecting the identity, culture and history of the region; and including indigenous peoples to promote their rights (Barents Cooperation 2017). The institution endorses a number of cross-border projects, the implementation of which promote confidence among the actors in the Barents Cooperation. A large number of the political priorities are realized through cross-border project implementation (Barentsinfo 2016a, b, c, d). Although BEAC and the Regional Council do not have any law-making power, they still produce results in the region through the measures taken by their working groups and via the implementation of cross-border projects. While BEAC does not have any concrete focus on or projects dealing directly with food security, the measures undertaken to promote the regional environment and the identity, culture and history of the region as well as particular concerns applicable to indigenous peoples and their inclusion within the cooperation framework offer implications that the body is relevant for addressing issues related to food security.

7.4.4   European Union (EU)

The EU is the governing institution for 28 European countries, known as members states (MS) (Europa 2017). The EU remains a prominent actor in the Arctic, with two member states and two other states tied to the EU through the EEA agreement. These states are Finland and Sweden and Iceland and Norway, respectively. Thus the EU-adopted legislations, in addition to member states, to a great extent also bind Iceland and Norway. There are two major sets of actions within the framework of the EU—regulations and directives. Regulations are those laws applicable to, and binding upon, all member states directly without any further actions to be taken at the national level (EU 2013). Directives are considered laws that bind the member states, or groups of member states, to achieve a particular objective. However, these must be transposed into national law to become effective (EU 2013). In addition, another set of actions are called "decisions," by which the specific EU laws are directed to individual or several member states and/or companies or private individuals, which are binding in their entirety (EU 2013). Finally, the EU also adopts recommendations and opinions that do not have any binding force. While within the Barents region Finland, Norway and Sweden are generally bound by EU laws, with Russia, the EU has a special bi-lateral and collaborative relationship in many areas through its external action service—for example, on the environment & economy, security & justice and research & education, including food and cultural aspects (EEAS 2016). In specific areas, such as in food safety, the national legislations of the Nordic EU member countries sometimes go above the standards set by the EU, which apparently better complements the governance of food security.

Nevertheless, the EU is believed to have some of the highest food safety standards in the world, mainly due to the solid set of legislations in place. In 2002, the European Parliament and the Council adopted Regulation No. 178, laying down the general principles and requirements of food law (General Food Law Regulation) (EC regulation 2002). This Regulation is the foundation for food and feed law, which sets out an overarching and coherent framework for the development of food and feed legislation covering the EU and its member states.

Furthermore, it lays down the general principles, requirements and procedures that underpin decision-making in matters of food and feed safety (EC regulation 2002). The EU "Food Law Regulation" ensures a high level of protection of human life and consumer interests in relation to food while at the same time ensuring the effective functioning of its internal market (EC regulation 2002). An important element of this Regulation is the responsibility of food and feed businesses to ensure that only safe food and feed is placed on the market. Food is deemed unsafe if it is potentially injurious to health, unfit for human consumption or contaminated in such a manner that it would be unfit for human consumption. The EU (in this case the European Commission) has gone above the set international guidelines to provide sufficient regulations and procedural measures on food. Also, according to a European Commission memo published in December 2012, around 98% of food legislation is harmonized at the EU level (Brans 2017).

Furthermore, the European Food Safety Authority (EFSA) was established in 2002 as an agency that communicates with member states by producing scientific opinions and advice on existing and emerging risks, which are used as the basis for forming European policies and legislation associated with food safety (EFSA 2016). In 2013, the EU set up a network called the "Food Fraud Network (FFN)" in response to the horse meat crisis. This network is aimed at allowing the EU countries to work in accordance with the rules laid down in articles 36–40 of the Official Controls Regulation (Regulation 882/2004, rules on administrative cooperation and assistance) in matters where the national authorities are confronted with possible intentional violations of food chain law with a cross-border impact. The FFN has 28 national food fraud contact points in member states, Norway, Iceland and the Commission. Overall, the EU legislations and measures in relation to food safety and security apply strictly to EU territories in the Barents region. Concerning the Russian Barents region, in addition to developing bi-lateral cooperation between the EU and Russia, the use of a forum such as BEAC could lead to the promotion of further collaboration between the EU and Russia, as the EU Commission is a member of BEAC.

7.5 National Measures

7.5.1 Finland

There are also national measures available to ensure the safety and security of food. In Finland, the administration of the food safety system is organized at four levels— the central administrative level, the ministerial level, the regional level and local levels (461 municipalities). In addition to compliance with national regulations as well as with the EU and other international regulations to which Finland is committed, various ministries assume responsibility for the development of strategies in relation to food safety and security. For example, the Ministry of Economic Affairs and Employment directs food control and quality matters related to all foods of non-animal origin as well as the market control of all foods (TEM 2017). The Ministry of Agriculture and Forestry takes overall responsibility for the control of the primary production and hygiene of foodstuffs of animal origin (MMM 2016). The Ministry of Social Affairs and Health deals mainly with the hygiene of foodstuffs of non-animal origin and the hygiene of all foodstuffs at the retail level and catering. Additionally, the Finnish Food Safety Authority, *Elintarviketurvallisuusvirasto* (Evira), aims at ensuring food safety, promoting animal health and welfare and developing the prerequisites for plant and animal production and plant health. Evira manages, directs and develops the control of products used in the primary production of foods and agriculture, the main goal of which is to ensure effective, efficient, consistent and risk-based targeted control across the entire food supply chain (Evira 2016). The practical enforcement of measures is carried out by local municipal authorities under the direction of provincial governments. Local municipal authorities take care of the control of the intra-community trade of foodstuffs of animal origin but only have power in their respective territories. Finland's food legislation is to a large extent harmonized with that of the EU, complementing many of the same rules and regulations. Finland endorsed its food act in 2006 (Wideback 2011). In 2015, a report entitled "Food Tourism Strategy for 2015–2020" (Havas et al. 2015) was endorsed with a view toward promoting tools for greater food security. These measures are precisely applicable to the Barents region of Finland.

7.5.2 Sweden

In Sweden, the National Food Administration (NFA) is an autonomous government agency that reports to the Ministry of Enterprise and Innovation and is the central administrative authority for matters concerning food (Sverige Regering 2015). The Swedish National Food Agency, *Livsmedelverket*, works toward healthy dietary habits, safe foods and fair practices in food trades through regulations, recommendations and communication (NFA 2015). The NFA issues food standards and other food regulations by carrying out supervision according to the Food Act (SFS 1971: 511).

It also leads and coordinates food control (FAOLEX-Sweden 2005). NFA additionally provides information on important matters concerning food and water. It takes an active part in fulfilling the objectives on diet and health set by the prevailing regulations. Sweden is divided into 21 counties and 289 municipalities. Food and water control at the local level is usually placed under the responsibility of the municipal Environment and Health Protection Committees. At regional levels, it is carried out by the county administration, while at the national level the NFA takes care of the control of food safety.

7.5.3   Norway

The Norwegian Food Safety Authority (NFSA), *Mattilsynet*, is a governmental body whose aim is, in compliance with the prevailing regulations on food safety and security, to ensure that food and drinking water are as safe and healthy as possible for consumers and to promote plant, fish and animal health. NFSA measures cover the ethical keeping of animals and encourage environmentally friendly production (NFSA 2016). As a member of the EEA, Norway commits to complying with EU regulations on food safety standards, labeling and traceability (Storting 2013). On December 19, 2003, Norway adopted the *Matloven Act* relating to food production and food safety (FAOLEX-Norway 2004). This Act was the result of merging 13 former acts that dealt with food safety, plant health and animal health. Norway, along with Sweden and Finland, apply stricter salmonella control and border control than other EU countries. The Norwegian Food Safety Authority's role is to draft and provide information on legislation; perform risk-based inspections; monitor food safety as well as plant, fish and animal health; and provide updates on developments in the field and plan for emergencies. NFSA also advises a number of Norwegian ministries, including agriculture and food, fisheries and coastal affairs and health and care services.

7.5.4   Russia

In Russia, while the *Duma* (the national legislative assembly) issues the legislative framework on food law and regulations, the Federal Veterinary and Phytosanitary Surveillance Service (known as *Rosselkhoznadzor*) under the Ministry of Agriculture monitors veterinary and phytosanitary conditions and enforces legal requirements for veterinary and plant health (Vanderberg et al. 2015). It also holds responsibility in regard to protection against plant and animal diseases (Vanderberg et al. 2015). In August 2012, Russia underwent some big policy adjustments after entering into the World Trade Organization as its 156th member. The country established the legal framework needed to comply with the WTO Sanitary and Phytosanitary (SPS) Agreement and undertook the relevant commitments on how it was to comply with the SPS Agreement and its other WTO commitments affecting trade in agricultural products (WTO-SPS 2015). The main changes related to the WTO accession of Russia concern market access improvements for goods and services (Vanderberg et al. 2015).

7.6　International Human Rights Framework

The mainstream structure of international human rights framework comprises three documents: the Universal Declaration of Human Rights adopted in 1948 and the two Covenants adopted in 1966—the ICCPR and ICESCR. These three documents are occasionally referred to as the international bill of rights (Perry 2005). As discussed earlier, the international human rights framework guarantees the right to food as it relates to the consumption, safety and availability of as well as access to food. The specific reference to the right to food is found in Article 11 of the ICESCR. In addition, several other articles within the human rights framework are connected in one way or another to food, such as the right to health or the right to enjoy a healthy environment (Valente 2014; Ayala and Meier 2017). Moreover, the right to food is not about physical sustenance only; the right as such is addressed as a form of enjoying a right to culture, too, in particular for traditional and indigenous communities.

Generally, culture consists of beliefs, practices and rituals held by a specific group of peoples living in a particular geographical setting, which are transmitted from one generation to the next. For certain communities, such as for indigenous communities, culture includes the maintenance of traditional knowledge and other ecological and local knowledge concerning land use management and biodiversity conservation. The preservation of a traditional subsistence economy, for example, in the context of indigenous peoples also forms part of their culture. The UN Special Rapporteur of the Sub-Commission on Prevention of Discrimination and Protection of Minorities, Francesco Capotorti, asserted that "culture" should be interpreted broadly to include customs, morals, traditions, rituals, types of housing and eating habits as well as the arts, music, cultural organizations, literature and education. The authoritative interpretation given by the Human Rights Committee (HRC), while interpreting Article 27 of the ICCPR in its General Comment No. 23, states that culture manifests itself in many forms, including a particular way of life associated with the use of land resources, especially in the case of indigenous peoples, which may include traditional activities such as fishing and/or hunting (General Comment 23 1994). The most recent General Comment produced by the Committee on Economic Social and Cultural Rights states that

> Culture, for the purpose of implementing article 15 (1) (a) [of the International Covenant on Economic Social and Cultural Rights] encompasses, inter alia, ways of life, language, oral and written literature, music and song, non-verbal communication, religion or belief systems, rites and ceremonies, sport and games, methods of production or technology, natural and man-made environments, food, clothing and shelter and the arts, customs and traditions

through which individuals, groups of individuals and communities express their humanity and the meaning they give to their existence, and build their world view representing their encounter with the external forces affecting their lives. (General Comment 21 2009)

The reference to the right to food from the viewpoint of culture is also addressed in the United Nations Declaration on the Rights of Indigenous Peoples (UNDRIP)—the Declaration basically highlighted the group component of rights, specifically as applicable to indigenous peoples (Wiessner 2009). This group component of rights, and particularly concerning indigenous peoples, is also presented in the ILO Convention No. 169 adopted in 1989 (UN IASG 2014). The particularly important aspects of the Convention are the commitments to recognition of the culture and cultural identity of indigenous peoples, their participatory rights, their right to be consulted in matters that affect their lands and livelihood practices and their right to decide the matters that are of priority for them (ILO C169 1989). Generally, the human rights framework perceives individual rights rather than group rights. However, because of the nature of "culture" as a holistic agenda, without reference to the group as a whole (or a particular identity held in common by the members of a community), the right as such cannot be meaningfully identified.

The "practice of culture" by an individual must conform to certain norms that the particular society prescribes or preserves. Moreover, an individual's enjoyment of culture not only includes an essential right to take part in cultural life but also includes rights to access and to contribute to the development and the formation of culture, making the culture a basis for the community identity. A view similar to this is found in the UNESCO report, which suggests that the right to culture is presented not only as a right to enjoy a way of life but also to enjoy cultural freedom as a collective freedom, referring to the right of a group or people to follow a way of life of their choice (UNESCO 1995). The practice of choice is clear among the indigenous communities who rely on traditional and nature-based resources as their food choice, which in fact promotes the concept of food sovereignty. Such a group component or collective dimension in the exercise of rights has been, and is, an integral part of understanding the right to culture as an entitlement.

The cultural significance of food, as shown above, has been strongly presented in the interpretation of human rights documents. For example, the HRC has interpreted Article 27 of the ICCPR rather broadly, highlighting the reference to food as an element of culture while addressing practices of hunting and fishing. In *Länsman et al. v. Finland*, the HRC clearly indicated that culture cannot be determined in "abstracto"; rather, significant impacts having an effect on the natural environment do have the potential to obstruct the enjoyment of culture in particular instances (Länsman et al. 1992). We have highlighted throughout this book that environmental elements play an integral role in the promotion of food safety and security. The countries within the Barents region are parties to all mainstream human rights documents. However, ILO Convention No. 169 is applicable only to Norway, and Russia has abstained in voting during the adoption of the UNDRIP. Yet, reference to these human rights documents sets standards in the governance of food security both as individual rights as well as group rights.

7.7 Concluding Remarks: An Assessment of Food Security Governance

Food security in the Arctic-Barents does not have any holistic governance frame work. Yet, the piecemeal approach applied to Arctic governance refers both explic itly and implicitly to food security (Young 2016). Food security has a number o aspects associated with it, such as nutritious values, safety issues, health compo nents, adequacy and accessibility. It also has cultural components. While specifi regulations addressing, for example, the impacts of climate change (which furthe lead to human activities such as oil and gas extraction, mining and tourism havin the potential to affect food security) apply in the context of food security, the inter national human rights framework also offers safeguards in the governance of food security. Although it is argued that solid regulation and good governance are crucia to ensure sustainability and human rights as elements of global food security an sustainable food systems (Lundqvist et al. 2015), we argue that the governance o food security needs to embrace regional characteristics both in regard to legislatio as well as policy documents. Support for such an approach can be found in the worl of Vapnek and Spreij, according to whom when developing a legal framework fo food security, all elements of food security policy and overall goals, both at th national and regional levels, are to be taken into account (Vapnek and Spreij 2005)

Given that the ability of communities to feed themselves depends partly or national and international food markets, the demand for land and the accessibility o natural resources (USAID 2013), addressing each of these challenges necessitates cross-sectoral response as it relates to the region. Regulations are put in place t minimize risk, covering risks in regard to food safety, population health, economi viability and the environment (Bloom et al. 2016). Because the number of risk associated with food, the food sector has become one of the most highly addresse sectors, with an increasing number of sporadic regulations (Bloom et al. 2016), and within a cross-sectoral structure, the governance of food security requires prope coordination. With climate change and human impacts looming over the region, it i vitally important to have the necessary policies and frameworks in place to dea with such threats and risks.

We suggest that regulation alone cannot ensure food security; the construction o strategies and their implementation through efficient and effective action plans adopted both at the regional and local levels in a coordinated framework are needed to guarantee better food security in the Barents region. The action plans, for exam ple, as argued by Rockström et al. (2017), are to be complemented with innovative financial mechanisms incentivizing carbon management in the food system. The authors called for agro-industries, farms and civil societies to develop a worldwide strategy for sustainable food systems to drive healthier, low-meat diets and reduce food waste. In regional settings, such strategies directed at promoting food security need to be multifaceted and respond to needs in a cross-sectoral manner, but in a coordinated form, to sustainably address all of the threats and risks.

In the Barents region sustainability is integrally connected to climate strategies, as illustrated in the beginning of this chapter. Although the goals of the Paris Agreement are aligned with science and can, in principle, be technically and economically achieved, alarming inconsistencies remain between science-based targets and national commitments (Rockström et al. 2017). This can be seen as a drawback in the Barents region, as Russia has not yet ratified the Paris Agreement. In addition, as discussed in this chapter, the prevailing legal instruments do not effectively cover all areas of the Barents region in a similar way. For example, the London Protocol, which entered into force in 2016, does not bind Finland and Russia. There are similar such drawbacks in other regulations, too.

However, the human rights framework is argued to be applicable, to address food security governance. It is often argued that human rights are universal, interdependent, indivisible and interrelated (Brems 2009). The right to food is connected to many other rights, such as the right to life, the right to health and, as discussed in this chapter, the right to the practice of one's own culture. The right to food cannot just be viewed as a mere entitlement; it is also about empowerment where the rightholders' choices are reflected. As we have argued food sovereignty as part of food security, in particular concerning indigenous peoples, such empowerment allows them to exercise their legitimate right to self-determination.

States' obligation to protect the right to food requires a responsive and efficient regulatory framework that clarifies the rights and obligations of rights-holders and duty-bearers and provides an enabling environment to implement the right to food (USAID 2013). Implementing such a framework in regard to the right to food must be complemented by accountability mechanisms. It is also argued that when regulations are founded on a human rights-based approach, it can significantly advance the objective of promoting food security (USAID 2013). In the Arctic-Barents region, the rights of indigenous peoples in relation to food as a cultural commodity is important. However, the human rights framework in connection to indigenous peoples' rights has its shortcomings. Firstly, mainstream human rights documents do not explicitly refer to group rights. Secondly, specific documents, such as the ILO Convention No. 169, has been poorly ratified—only Norway in the region is a party—and UNDRIP, despite its normative importance, provides nonbinding obligation, and Russia even abstained from voting in the process of its adoption. Therefore, we argue that better food security governance in the regional context requires adherence to these treaty obligations, too.

Chapter 8
Knowledge Gaps and Recommendations: An Analysis

It is evident from the preceding seven chapters of this book and from other cited works that there are gaps in the knowledge about the current state of food security. In addition, the existing laws and regulations and other policy instruments are enacted in a fragmented manner, offering a clear lack of coherence in governance approach. We, therefore, while highlighting the existing governance tools, argue for better coordination in governance approach, particularly as it relates to the region's food security governance.

The big concern in terms of food supply is about meeting the need for the expected global population of 9.8 billion by 2050 (UN 2017). This global grand challenge resonates with Robert Malthus's theory of population and food supply of more than 250 years ago, which is still very relevant today (Malthus 1826). As the global population increases, there will be more in need of shelter and more mouths to feed. All these will put a strain on the available resources, which necessitates the need to extract more natural resources from land and water. As a result, industries are constantly on the lookout for opportunities to meet these needs. Apart from the challenges and opportunities that will result through shipping along the Northern Sea Route, the Arctic-Barents region—as a last frontier for the provision of oil, gas, timber and also non-wood forest products—is becoming highly relevant. The world's largest certified non-agricultural organic area is also to be found in this region. Therefore, one overarching question will be how best to maintain a good balance between the extractive industries and the ecology of the region in a sustainable manner given that the latter is connected to safe and secure food production.

As stated in the introduction (Chap. 1), in relation to the concept of food sovereignty and how it might be more appropriate in the promotion of food security in the Barents regional context, we (quite contrary to the remarks by Gordillo and Jeronimo (2013) on why food sovereignty and food security are antagonistic to a modern state) argued that the promotion of greater say and participation in decision-making for communities is today recognized as a basic democratic function, which does not in any way jeopardize the notion of state sovereignty. As rightly pointed out by Kuhnlein and Burlingame (2013), and discussed in this book (under Sect. 4.1.

© Springer International Publishing AG, part of Springer Nature 2018   111
K. Hossain et al., *Food Security Governance in the Arctic-Barents Region*,
https://doi.org/10.1007/978-3-319-75756-8_8

Imported versus traditional foods), the foods that are purchased from markets in the region often come through globalized industrial outlets and, as such, are also part of the peoples' food system. Assaults on "indigeneity" and self-determination were observed to be linked to the disparity in food security, health effects, poverty, education, nutrition, household crowding and poor access to and utilization of health care (Kuhnlein and Burlingame 2013).

This calls for means by which local people are empowered in making a difference regarding food sovereignty. Food sovereignty was recognized by the Nyéléni Declaration as responding to the ecologically appropriate production, distribution and consumption of food as well as social-economic justice by using local food systems as ways to tackle hunger and poverty, which will guarantee sustainable food security for all peoples (Barkin 2016). The Declaration advocates for trade and investment that will serve the collective aspirations of a society by promoting community control of productive resources; agrarian reform and tenure security for small-scale producers; agro-ecology; biodiversity; local knowledge; the rights of peasants, women, indigenous peoples and workers; social protection and climate justice (Nyéléni Newsletter 2013).

A comprehensive review of the evolution of the use of these concepts in the academic literature is available (e.g., in Edelman 2014; FPH 2011). Food sovereignty highlights the relationship between the importation of cheap food and the weakening of local agricultural production and populations. Both food security and food sovereignty emphasize the need to increase food production and productivity to meet future demand. Both concepts stress that the central problem today is access to food and thus involves redistributive public policies in terms of income and employment. Food security does not take into account the concentration of economic power in the different links of the food chain and in the international food trade or the ownership of key means of production such as land or, more contemporarily, access to information. On the other hand, the concept of food sovereignty begins precisely with noting the asymmetry of power in the various markets and the different spheres of power involved in food production as well as in the areas of multilateral trade negotiations (Gordillo and Jeronimo 2013). Hence, the authors have called for democratic states to balance these inequalities and consider food to be more than a mere commodity.

Food security also differs from food sovereignty in terms of how food is produced. In the document prepared by Gordillo and Jeronimo (2013) for the FAO, they noted that food security relies on so-called industrial agriculture, based on the intensive use of fossil fuels; biological agriculture, which uses biomass and biotechnologies, of which GMOs are only a part; and organic agriculture, which involves processes that require various forms of certification. In contrast, the concept of food sovereignty is clearly focused primarily on small-scale agriculture (including livestock, forestry and fisheries) of a non-industrial nature, preferably organic, mainly using the concept of agro-ecology (Gordillo and Jeronimo 2013).

In regard to food security in the Arctic-Barents region, an understanding of how these changes will affect the traditional foods that have nourished the peoples of the region for generations requires more research. As discussed earlier (Chap. 4), the available foods in the region are being affected by human activities, globalization and climate change (Chap. 6) in an unprecedented way. It will therefore be important to identify knowledge gaps to ensure that these foods will continue to sustain and meet the nutritional needs of the people as well as maintain the cultural affinity related to traditional foods (Chap. 5).

At the other end of the spectrum are issues related to the safety of these traditional foods and water security in the region. The extraction of solid minerals, oil and gas in the region and their impacts on these foods as discussed in Chap. 6 are areas of concern that require coordination of the existing legal tools.

This book has been completed by fully analyzing and articulating the published research and through conducting interviews with researchers who are very active and continuously working on the ground with locals all throughout the Barents region. Due to the identified infrastructural problems in Russia, the authors also agree with the remarks of Rautio et al. (2013) that the future of foreign business in the Russian parts of the Barents region will be closely linked to the economic and political development of Russia and its northwestern regions. Food security issues will be affected by the global demand for the natural resources that are available in the region and the long-term effects of climate change. Also in the near future, the Arctic policies that are currently being developed, such as by the EU and China, will likely play a role in food security and governance in the region.

8.1 Research Gaps

First, food security in the Barents region has been relatively under-researched, as the comprehensive food security research that was conducted 15 years ago by Duhaime and Bernard (2001) is still being utilized today. The authors suggested that there is a lack of information on the current state of food security in the region. This is frequently noticed when analyzing and comparing food security issues in the Barents region and the entire Arctic. The Barents region is largely understudied and has little documentation. The lack of documentation and observation on food security makes it difficult for policy makers to assess and address the needs of the indigenous and non-indigenous peoples of the region without a clear understanding of the challenges that are affecting the communities. One such challenge is the case of fisheries in the region due to a decline in qualified scientific experts, which has made unresolved taxonomic issues regarded a problem in many taxa at many levels (Reist et al. 2013). It has been observed that anthropogenic factors can either affect diversity directly (e.g., specific taxa or forms exploited in fisheries) or indirectly (e.g., climate change affecting the productivity of water bodies) by altering the processes through which diversity is maintained (ABA 2013).

Duhaime and Bernard (2001) further noted that there are scarce data available on the food consumption patterns in the Euro-Arctic-Barents region; this is the same problem that we encountered and also a key gap in this book. Data on food consumption patterns, especially among traditional and imported food products, are important for determining their exact health consequences, contaminant intake and other factors that result from both traditional and market-based foods. The authors also specified that food security in the Euro-Arctic-Barents is subject to a plurality of conditions attributable to the heterogeneity of the territory and national contexts that influence the governance of these territories. The inequality of situations makes it hard to generalize findings to the entire region and calls for caution in the assessment that may be made when examining food security in the communities (Duhaime and Bernard 2001). In compiling this book, we had to rely on generalization within the Arctic and the homogenization of information in the Barents region due to a lack of specific data on certain issues or communities within the Barents region. For example, statistical data on certain foods, their consumption and value addition, especially in Northern Russia, are hard to obtain. Similarly, specific data for these indicators for indigenous and non-indigenous groups within the region are difficult to obtain.

In addition, Duhaime and Bernard (2001) suggested that certain scenarios should be developed to provide decision makers with the information and data required to orient their decisions toward the sustainability of economic systems, the sustainable exploitation of resources, the optimization of social health conditions and the preservation of food security of communities.

Much work must still be done, and the fact that we are not yet able to propose scenarios for optimizing food security is an important limitation of this volume (Duhaime and Bernard 2001). We draw the same conclusions concerning a need for further research on food security scenarios, as it is difficult to build scenarios without further research and knowledge of the Barents region. This is necessary to address policy makers and other stakeholders in creating a sustainable strategy for food security in the region. Further research is an important aspect of food security in the region, especially as we have identified, along with other researchers, the numerous effects of climate change and human activities that are affecting food security in the region. With more research, we believe policy makers can then better assess and implement effective strategies, policies and procedures for further dealing with the threats and challenges to food security that individuals and communities are facing in the region. The exposure to various chemicals used for varied purposes are of emerging concern in the Arctic, which should be considered as potential elements for future research or monitoring and possibly for consideration under relevant global and/or regional regulations. In addition, these chemicals of emerging concern contribute to a broader understanding of how Arctic pollution is changing.

8.2   Governance

8.2.1   Organizations

As assessed above, there are a number of institutions that regularly contribute to and work toward improving the Arctic and the Barents region. These institutions were discussed in Chap. 7; notably among them are the Northern Dimension, the Barents

Euro-Arctic Council and the Arctic Council, and they contribute in some ways to the Barents region. Their work focuses on a number of topics, such as the environment, indigenous peoples, the economy and culture. Despite the wide range of work that these organizations do, much of their work is limited concerning food security in the region. We have pointed out that they do deal indirectly with such issues through adopting policy measures in relation to environmental and economic transformation and, most importantly, in connection to concerns arising out of climate change and associated consequences. However, they do not comprehensively address numerous factors pertaining to food security. Furthermore, the Arctic Council and the Barents Euro-Arctic Council both have indigenous peoples' working groups, whereas the Northern Dimension does not directly involve indigenous peoples. This becomes limiting when working in a region that has diverse groups of indigenous peoples that rely to a large degree on traditional food. More importantly, these organizations do provide policy documents; rather they have a status of only "soft-law." Thus, these are merely equivalent to making recommendations, and therefore, these strategies are not binding on the states in the region.

On a more positive note, these organizations have provided a number of reports and completed various research projects in the Barents region that are directly and indirectly related to food security. Some examples produced by the Arctic Council and its working groups are found in the Arctic Human Development reports and the AMAP Arctic Assessment reports dealing with a number of issues, which we have provided as the basis of references for this book—such as issues related to human health, the effects of climate change and the adverse impacts on the natural and human environment. One such report, "Food and Water Security Indicators in an Arctic Health Context," was published during the Arctic Council Swedish chairmanship in 2013 (Nilsson and Evengård 2013). This was the first report published by the Arctic Council that addressed food security across the Arctic with specific reference to the Barents region. However, it was also noticeable that the report gathered much data from the North American Arctic, where much of the research on the Arctic has been completed, further solidifying the point that there is still much more research to do in the Barents region. The report also explained that many indicators of relevance from an indigenous Arctic-Barents perspective, such as non-monetary food accessibility and conditions for hunting/fishing/collecting/herding, had not been repeatedly monitored before and were deemed in need of methodological development in the future (Nilsson and Evengård 2013). Despite the lack of comprehensive data from the Barents region, the report is important for assessing and determining the indicators for both water and food security, which is relevant across the whole Arctic. While these indicators were a priority in the first phase of the project, they also make additional recommendations to continue this work in an international study but within the national and regional context. Lastly, due to all the changes in the Arctic-Barents region (climate change and increase in human activities), Nilsson and Evengård suggested further international cooperation using such indicators for the surveillance of food and water security in certain nations and regions in the Arctic. Moreover, such indicators would be useful in the Barent region to further determine the food security situation in the region.

The work of these institutions continues to be extremely important in the Arctic Barents for further strategy, research and soft-law development. They effectivel utilize their working groups with the involvement of indigenous peoples from al across the region to learn about the challenges to, for example, the economy, envi ronment and culture. However, in the wake of climate change and human challenge in the Barents region, these institutions should put more effort into the challenge o food security instead of merely addressing it indirectly through related projects Furthermore, organizations such as the Arctic Council, Northern Dimensions an BEAC should place priority on strategies, recommendations and solutions in regar to solving food security issues in the region. We do not believe that there is a lack o governance structure in the region per se but rather a lack of direct focus and coor dination among the various tools and mechanisms regarding food security issue within these institutions.

8.2.2 States

States in the Barents region play a large role in promoting and protecting food secu rity and food safety among their populations. This can be done through trade rela tions, international agreements, laws, policies and the recognition of certain right within the Barents' own territory. Trade relations and political relations within the Barents region have been hampered in recent years with drastic effects on the econ omy because of the Crimean crisis, which started in March 2014. This has caused a price hike of food in many areas within the region and with some limitations as to what kinds of, and how much, food is available in the stores. For those who depen on market-based and imported food, the inability to afford such food may become challenging and lead to further food insecurity. Furthermore, the sale and trade o some foods have been banned from entering the countries—for example, the trad of dairy products between Finland and Russia as well as the trade of fish betweer Norway and Russia. This has had impacts in each country and especially in the northern regions of the Barents region. For those individuals and communities, both indigenous and non-indigenous, this can have effects on their food security situa tion as well. Furthermore, individuals and communities cannot sell local or tradi tional foods to markets or might need to seek new markets, which could become a challenge. Therefore, political and economic relations between countries in the region are important for the political, economic, food and human security of the individuals and communities.

The ratification of certain international agreements is important for achieving food security in the Barents region. There are a number of international agree ments mentioned in previous chapters that are of vital importance for food secu rity, but they are indirectly related, focusing on key areas such as shipping, air pollution, marine pollution, biodiversity, contamination from various sources, the effect of climate change and many more. It is, however, important to commit to such regulations to protect the environment and environmental security, which eventually serve individuals and communities living in the region. For example, much work needs to be done on climate change efforts in these countries. The signing and ratification of the new Paris Agreement and the funding and implementation of regional climate strategies in all areas of the Barents region would certainly promote food security. On several levels, the food industry is one of the most regulated industries in the world. Laws, regulations, policies and strategies are aimed at ensuring food security and safety. Generally, at the national level, states play a large role in the enforcement of these regulations by adopting effective measures through which they adapt to the changing circumstances, for example, around the threats of climate change and the other associated factors, such as human activities. Therefore, we argue that such agreements and strategies need to be better coordinated to strengthen food security and safety, which will eventually ensure overall human security in the region.

8.2.3 Human Rights Framework

To the extent that food security is connected to the human rights framework, as referred to in the previous chapter, we highlight that food is one of the fundamental human rights linking not only individuals' right to life but also the right to culture for certain ethnic and cultural communities. The right to food, including the right to adequate food, has been deemed one of the most important rights in addressing and further promoting food security. This right has been endorsed as a supreme human right in the most crucial international and national legislation around the world. It was argued that without the right to food one cannot guarantee life, dignity or the enjoyment of other human rights. In response to this claim it was necessary to achieve a better definition of the concept of the right to food that will foster the creation of concrete tools to improve its implementation; this led to the adoption of General Comment No. 12 in 1999 by the Committee on Economic, Social and Cultural Rights (CESCR) in relation to the right to adequate food (Gordillo and Jeronimo 2013). The document emphasized that the right to adequate food implies the right to food in quantity and quality sufficient to satisfy the dietary needs of individuals and the right to food that is free from adverse substances and acceptable within a given culture as well as sustainable access to this food. Despite the importance of General Comment No. 12, it is not mandatory and has not been endorsed by all governments (General Comment No. 12 1999). Solidifying the right to food in the constitutions of those countries in the Barents region has not gained much traction. In fact, Finland is the only country in the region that has recognized the right implicitly. We therefore argue that countries in the Arctic-Barents region would benefit from promoting the human right to food security by endorsing the guidelines elaborated in General Comment No. 12.

8.2.4   Land Rights

Land and land rights are important for both indigenous and non-indigenous peoples surrounding further food security. Much of the traditional and local food sources in the Arctic-Barents region come from the land; therefore, access to, control of and rights over lands are important in this regard. In fact, securing rights to land are a critical, but often overlooked, factor in achieving household food security and improved nutritional status. Furthermore, the Office of the High Commissioner for Human Rights explained that, to produce his or her own food, a person needs land, seeds, water and other resources, and to buy food, one needs money and access to the market. In the Barents region many of the discussions surrounding land rights, especially among indigenous peoples, revolve around ILO Convention No. 169. Such rights would enable people in the region to take broader rights on lands through managing forests, taking care of the environment and making key decisions as collectives. Moreover, such rights will help them to exercise control over their own economic, social and cultural development. This will further enable them to participate in the formulation, implementation and evaluation of plans and programs for their own development. ILO No. 169 has been signed and ratified by Norway, but Finland, Sweden and Russia have still failed to recognize and commit to the Convention, hence the right as such is not guaranteed for the indigenous peoples of these countries. We argue that such ratification and acceptance of ILO No. 169 would help the indigenous peoples of these countries to better organize their lands for the promotion of food security.

8.2.5   Collective Rights

Group rights, in particular concerning the groups of indigenous peoples, are significant for communities who participate in food-related activities together and on a regular basis, such as the production, gathering, hunting, trade and consumption of food. The right is exercised, in addition to consumption, also as a cultural right. The right to culture cannot be practiced alone in an isolated manner. Individuals, in community with the other members of the same group, enjoy their right to culture. Therefore, group rights and collective rights must also be recognized and proclaimed, as the hunting and fishing rights utilized by many groups and communities in the region are important for ensuring further food security, both for physical consumption and for protecting cultural rights. The rights allowing access to and the utilization of traditional and local foods are also crucial for the sustainability of a population and even more so in the Arctic-Barents environment.

8.3 Recommendations and Future Outlooks

Food security and its governance in the Arctic-Barents region is at a critical point in time, where the threats and challenges to the people living there are much higher than ever before. Climate change, human activities and globalization are constantly having impacts on both the indigenous and non-indigenous ways of life in the Barents region. These threats have negative effects on the four pillars of food security: the accessibility, availability and utilization of such food and the overall food system's stability (Chap. 6). The extent of these threats differs between the four countries of the region given that the national policies in these four counties are different. However, the threats are far-reaching. They extend across borders. The effects not only pose threats to the food security of the people but also influence all other aspects of human security, too, in that environmental, community, economic, personal, political and health security are all in one way or another connected to food security. Strengthening food security will also strengthen these human security aspects for the individuals, groups and communities located in the area. Food security can be strengthened through effective policies, strategies and legal tools as discussed in relation to organizations, states and international actors that work in, and alongside, the region. However, primarily, there is a need for more research to be carried out in the Barents region that determines the exact food security situation for specific areas to create effective strategies and policies that target the identified obstacles. The relevant data obtained from such research will help to harmonize and enhance the food security in the region.

The Arctic-Barents region has recently been recognized as an important expanse with great potential in terms of food production in the future. The region has traditionally focused on ocean and sub-sea ocean resources, but there is also potential in increased terrestrial food production. The Arctic-Barents can help to meet the demands of terrestrial food production, since this region faces less risk of water scarcity, and the growing season will be extended with climate change (Bardalen 2016) given that an effective governance approach to food security is in place. Therefore, there is a need for research as well as extension services to disseminate new knowledge and best practices, among others, to those working the farmland in the region (Bardalen 2016). The need to explore this potential and increase food security through collaborative efforts is gaining attention in regard to identifying the conditions for increased production, which will improve food security in the northern regions and increase the added value of food originating in the Arctic that will be meant for both local and southern markets. Nowadays, with increased problems in terms of food adulteration and fraud, consumers are keen to know the geographical origin of their foods, thereby encouraging the traceability of foods from suppliers with added value.

If food sovereignty were to be recognized, the world would look to the North for leadership and direction in a changing Arctic. This would mean not abandoning new development but rather working with northerners to set the vision and agenda. At the local scale, one practical way to move in this direction would be to facilitate more meaningful engagement in, and even leadership of, the environmental assessment (EA) process for all new development (Noble and Hanna 2015). Explicitly incorporating local people's priorities, such as food security and health into EAs, and elevating these local needs over simple profits, would be an enormous first step. Food sovereignty goes beyond the ordinary meaning of food security, and but in this study we integrate it in the context of the Arctic-Barents region. The indigenous peoples and cultures indicate how other forms of knowledge acquisition and dissemination are relevant for governance and management processes in the region's marine environment. Linking development to security and human rights is a strong argument for food security. This can be realized by assuming the autonomy of actors to define their own food policies through food sovereignty and emphasizing that the concept of food sovereignty is not antagonistic toward or conflictive with the concept of food security. An integration of indigenous and local knowledge within the various fields of the natural and social sciences to effectively inform management processes will require considerable ongoing efforts (PAME 2013). Reconnecting indigenous peoples with their traditional territories and reversing some of the restrictive regulations against historical hunting and the harvesting of plants were also identified (Laird 2002; UNPFII 2009; CBD 2010) as ways that may help to restore and maintain traditional resources for indigenous peoples. States should work more closely with the Arctic-Barents residents to identify and promote effective models for enabling the inclusion of traditional knowledge related to food security into decision-making processes for a sustainable resource management strategy in the region.

We highlight the relevance of small and medium-sized enterprises that are making efforts to add value to these food systems with technological inputs. Such initiatives and other related innovations will help to provide jobs as well as help in contributing to the traditional processing, packaging and marketing of these foods and, in this regard, to relevant best-use technology, such as the promotion of digitalization. Overall, the sharing of best practices and strengthening the governance related to food security will help promote food security and launch innovative products from the traditional foods available in the Arctic-Barents region.

References

ABA (2013) Arctic biodiversity assessment, Chapter 6. Fishes. Conservation of Arctic flora and fauna (CAFF www.caff.is). Accessed 14 May 2017

ABC News (2017) Putin signs decree to extend ban on Western food imports. http://abcnews.go.com/International/wireStory/putin-signs-decree-extend-ban-western-food-imports-48371855. Accessed 02 July 2017

ACIA (2005) Arctic climate impact assessment. Overview report. Cambridge University Press, Cambridge

Adams M (2017) Wild tradition: hunting and nature in regional Sweden and Australia. In: Head L, Saltzman K, Setten G, Stendeke M (eds) Nature, temporality and environmental management: Scandinavian and Australian perspectives on people and landscapes. Routledge, London

AHDR II (2015) In: Larsen JN, Fondahl G (eds) Arctic human development report: regional processes and global linkages. Nordic Council of Ministers, Copenhagen

Akram-Lodhi AH (2009) World food security: a history since 1945. Rev Can Etudes Dev 28:3–4

Akvaplan-Niva (2007) Climate and ecosystems. Arctic Research and Development. http://www.akvaplan.niva.no/en/arctic_rd/climate_ecosystems. Accessed 12 Aug 2017

AMAP (1998) AMAP assessment report: Arctic pollution issues. Arctic Monitoring and Assessment Program (AMAP), ISBN 82-7655-061-4, Oslo, Norway, pp xii + 859

AMAP (2009) Arctic pollution 2009. Arctic Monitoring and Assessment Programme, Oslo, p 83

AMAP (2011a) Snow, water, ice and permafrost in the Arctic (SWIPA): climate change and the cryosphere. Arctic monitoring and assessment program, Oslo

AMAP (2011b) Arctic monitoring and assessment programme. Assessment 2011: mercury in the Arctic. Executive summary and key recommendations. Arctic monitoring and assessment programme (AMAP), Oslo, Norway

AMAP (2014) Trends in Stockholm convention persistent organic pollutants (POPs) in Arctic air, human media and biota. In: Wilson S, Hung H, Katsoyiannis A, Kong D, van Oostdam J, Riget F, Bignert A (eds) AMAP technical report no. 7. Arctic Monitoring and Asessment Programme (AMAP), Oslo, p 54

AMAP (2015) Arctic monitoring and assessment programme. AMAP assessment 2015: human health in the Arctic. Oslo, Norway, 2015

AMAP (2016) Chemicals of emerging Arctic concern (CEAC) Assessment summary for policymakers. www.amap.no. Accessed 6 May 2017

AMAP (2017a) Adaptation actions for a changing Arctic (AACA) - Barents area overview report. Association of world reindeer herders, "Saami & Finns – Finland." Reindeer Herding: a virtual

© Springer International Publishing AG, part of Springer Nature 2018   121
K. Hossain et al., *Food Security Governance in the Arctic-Barents Region*,
https://doi.org/10.1007/978-3-319-75756-8

guide to reindeer and the people who herd them. http://reindeerherding.org/herders/Saami-finns-finland/. Accessed 1 April 2017

AMAP (2017b) chemicals of emerging Arctic concerns. Summary for policy makers. Arctic monitoring and assessment program report, April 2017. https://www.scribd.com/document/346684885/Chemicals-of-Emerging-Arctic-Concern#from_embed. Accessed 12 May 2017

AMSA (2009) Arctic Marine Shipping Assessment report 2009. https://cil.nus.edu.sg/wp/wp-content/uploads/2014/06/2009-Arctic-Marine-Shipping.pdf. Accessed 10 Jul 2017

AOOGG (2009) Arctic council, Arctic offshore oil and gas guidelines. http://www.pame.is/images/PAME_NEW/Oil%20and%20Gas/Arctic-Guidelines-2009-13th-Mar2009.pdf. Accessed 14 July 2017

Archibald CP, Kosatsky T (1991) Public health response to an identified environmental toxin: managing risks to the James Bay Cree related to cadmium in caribou and moose. Can J Public Health 82(1):22–26

Arctic Council (2009) Arctic-Barents marine shipping assessment: scenarios, futures, and regional futures to 2020. Arctic monitoring and assessment programme, AMAP assessment 2009: human health in the Arctic. Oslo, Norway

Arctic Council (2016) Frequently asked questions, November, 4, 2016. http://www.Arctic-Barents-council.org/index.php/en/about-us/Arctic-council/faq. 14 November 2016

Arctic flavours association (2017) Arctic Flavours 2017. Arktiset Aromit, Suomussalmi, Finland. http://www.arktisetaromit.fi/en/association/. Accessed 20 Jun 2017

Arctic Now (2017) Industrial poisons in the Arctic environment. https://www.arcticnow.com/science/2017/05/01/new-industrial-poisons-may-threaten-arctic-environment-says-study/. Accessed 3 May 2017

Arctic Portal (2017) http://arcticportal.org/arctic-governance/arctic-cooperation. Accessed 14 July 2017

Arlinghaus R, Tillner R, Bork M (2015) Explaining participation rates in recreational fishing across industrialised countries. Fisheries Management and Ecology 22:45–55

Austvik OG, Moe A (2016) Oil and gas extraction in the Barents region. In: Olsson MO (ed) The Encyclopedia of the Barents Region II, Interreg Project. Pax Forlag, Oslo, pp 115–121. ISBN: 9788253038780. Oil and Gas Extraction in the Barents Region. (PDF Download Available). Available from: https://www.researchgate.net/publication/314044405_Oil_and_Gas_Extraction_in_the_Barents_Region. Accessed 12 Jul 2017

Ayala A, Meier BM (2017) A human rights approach to the health implications of food and nutrition insecurity. Public Health Rev 38(1):10

AWRH (2016) "Sami & Finns–Finland." Reindeer Herding: a virtual guide to reindeer and the people who herd them. http://reindeerherding.org/herders/sami-finns-finland/. Accessed 29 Jun 2017

Baldursson S (2003) Module 10: living terrestrial resources of the Arctic and their use. Bachelor of circumpolar studies (BCS 311). University of the Arctic. http://members.uarctic.org/participate/circumpolar-studies/course-materials/bcs-311-land-and-environment-i/. Accessed 14 April 2017

Barański M, Srednicka-Tober D, Volakakis N, Seal C, Sanderson R, Stewart GB, Benbrook C, Biavati B, Markellou E, Giotis C, Gromadzka-Ostrowska J, Rembiałkowska E, Skwarło-Sońta K, Tahvonen R, Janovská D, Niggli U, Nicot P, Leifert C (2014) Higher antioxidant and lower cadmium concentrations and lower incidence of pesticide residues in organically grown crops: a systematic literature review and meta-analyses. Br J Nutr 112:794–811

Bardalen A (2016) Arctic agriculture: producing more food in the north. Arctic deeply, 2 November 2016

Barents Cooperation (2017) Barents Euro-Arctic cooperation. http://www.barentscooperation.org/en/About/Members. Accessed 26 July 2017

Barents Info (2016a) "Veps and Karelians." BarentsInfo.org. http://www.barentsinfo.org/Contents/Indigenous-people/Veps. Accessed 6 Nov 2016

Barents Info (2016b) The Barents Euro-Arctic region: cooperation and visions of the north. The Barents Euro-Arctic Region. The Nenets. BarentsInfo.org. http://www.barentsinfo.org/ Contents/Indigenous-people/Nenets. Accessed 6 Nov 2016

Barents Info (2016c) "The Saami." BarentsInfo.org. http://www.barentsinfo.org/Contents/ Indigenous-people/Saami. Accessed 6 Nov 2016

Barents Info (2016d) "Pomors". http://www.barentsinfo.org/Contents/Indigenous-people/Pomors. Accessed 6 Nov 2016

Barents Watch (2007) Climate in change: nature and society challenges for the Barents region. http://www.bioforsk.no/ikbViewer/Content/96985/BW07_engelsk_nett%20(2).pdf. Accessed 8 July 2017

Barkin D (2016) Food sovereignty: a strategy for environmental justice. World economics association conference: FOOD AND JUSTICE: ideas for a new global food agenda? http://foodandjustice2016.weaconferences.net/. Accessed 18 June 2017

Barrie LA, Gregor D, Hargrave B, Lake R, Muir D, Shearer R et al (1992) Arctic contaminants: sources, occurrence and pathways. Sci Total Environ 122(1-2):1–74

Bartsch A, Kumpula T, Forbes BC, Stammler F (2010) Detection of snow surface thawing and refreezing using QuikSCAT: implications for reindeer herding. Ecol Appl 20:2346–2358. https://doi.org/10.1890/09-1927.1

BBC (2014) "Nenets." October 29, 2014. http://www.bbc.co.uk/tribe/tribes/nenets/. Accessed 7 Nov 2016

BEAC (2016a) Barentsinfo.org. "Facts." http://www.barentsinfo.org/Barents-region. Accessed 26 Aug 2016. "Barents Sea – Physical characteristics." BarentsInfo.org. http://www.barentsinfo. org/Contents/Nature/Barents-Sea/Physical-characteristics. Accessed 10 Nov 2016

BEAC (2016b) Barents Euro-Arctic Council. "Geography." http://www.beac.st/en/About/ Barentsregion/geography. Accessed 10 Nov 2016

BEAC (2016c) Barents Euro-Arctic Council. "History." http://www.beac.st/en/About/Barentsregion/history. Accessed 10 Nov 2016

Beach H (1981) Reindeer-herd management in transition: the case of Tuorpon Saameby in Northern Sweden, p 542. (Doctoral dissertation, Acta Universitatis Upsaliensis). Uppsala, Sweden

Becker W, Pearson M (2002) Riksmaten 1997–98. Dietary habits and nutrient intake in Sweden. The National Food administration, Uppsala, Sweden, pp 1997–1998

Berg E (2014) Saami traditions: Márkomeannu's contribution to the revitalization of Saami food traditions. The Arctic University of Norway, November, 2014

Bersaglieri T, Sabeti PC, Patterson N, Vanderploeg T, Schaffner SF, Drake JA et al (2004) Genetic signatures of strong recent positive selection at the lactase gene. Am J Hum Genet 74:1111–1120

Berti PR, Receveur O, Chan HM, Kuhnlein HV (1998) Dietary exposure to chemical contaminants from traditional food among adult Dene/Metis in the western Northwest Territories, Canada. Environ Res 76(2):131–142

Bertozzi L (1998) Tipicidad alimentaria y dieta mediterra´nea. In: Medina A, Medina F, Colesanti G (eds) El color de la alimentacio´n mediterra´nea. Elementos sensoriales y culturales de la nutricio´n, Barcelona: Icaria, pp 15–41

Bessiere J, Tibere L (2013) Traditional food and tourism: French tourist experience and food heritage in rural spaces. J Sci Food Agric 93(14):3420–3425

Bjermo H, Darnerud PO, Lignell S, Pearson M, Rantakokko P, Nälsén C et al (2013) Fish intake and breastfeeding time are associated with serum concentrations of organochlorines in a Swedish population. Environ Int 51:88–96

Bjørklund I (2004) Sami pastoral society in northern Norway: the national integration of an indigenous management system. In: Anderson D, Nuttall M (eds) Cultivating arctic landscapes. Berghahn Books, New York, NY, pp 124–135

Bloom M, Grant M, Slater B (2016) Governing food: policies, laws, and regulations for food in Canada. The Conference Board of Canada. http://www.conferenceboard.ca/cfic/research/2011/ governingfood.aspx. Accessed 15 Nov 2016

Bogdanov EV, Ungureanu TN, Buzinov RV, Gudkov AB (2011) Morbidity of asthma population of the population in Arkhangelsk region. Human Ecol 12:8–13

Borgå K, Gabrielsen GW, Skaare JU (2000) Biomagnification of organochlorines along a Barents sea food chain. Environ Pollut 113:187–198

Brans H (2017) Food and agricultural import regulations and standards. http://agriexchange. apeda.gov.in/IR_Standards/Import_Regulation/Food%20and%20Agricultural%20Import%20 Regulations%20and%20StandardsBrussels%20USEUBelgiumLuxembourg12192016.pdf. Accessed 25 Aug 2017

Braun J (2011) EU energy policy under the treaty of Lisbon rules: between a new policy and business as usual (February 24, 2011). EPIN working paper No. 31. Available at SSRN: https:// ssrn.com/abstract=2001357. Accessed 12 July 2017

Brems E (2009) Human rights: universality and diversity, Kluwer Law International 2001, p. 20

Britannica (1998) The Editors of Encyclopaedia Britannica. https://www.britannica.com/topic/ Komi-people. Accessed 30 Jul 2017

Brown G (2016) The universal declaration of human rights in the 21st century: a living document in a changing world. Open Book Publishers. ISBN 978-1-783-74218-9

Brubaker M, Bell J, Rolin A (2009) Climate change effects on traditional Inupiat food cellars. Centre for Climate and Health, CCH Bulletin no 1, October 19, 2009. http://www.north-slope. org/assets/images/uploads/CCH-Bulletin-No-01-Permafrost-and-Underground-Food-Cellars-Revised-Final.pdf. Accessed 27 May 2017

Brustad M, Parr CL, Melhus M, Lund E (2008) Dietary patterns in the population living in the Sámi core areas of Norway – the SAAMINOR study. Int J Circumpolar Health 67:82–96

Bultrini D (2009) Guide on Legislating for the Right to Food. Food and Agriculture Organization of the United Nations, Rome, 2009

Burek KA, Frances MDG, O'Hara TM (2008) Effects of climate change on Arctic-Barents marine mammal health. Ecol Appl 18(sp2):S126–S134

Burke M, Lobell D (2010) Climate effects on food security: an overview. In: Lobell D, Burke M (eds) Climate change and food security: adopting agriculture to a warmer world. Springer, Rotterdam, The Netherlands, pp 13–30

BWM (2004) International convention for the control and management of ships' ballast water and sediments. http://www.imo.org/en/About/Conventions/ListOfConventions/Pages/ International-Convention-for-the-Control-and-Management-of-Ships%27-Ballast-Water-and-Sediments-(BWM).aspx. Accessed 28 July 2017

Caldwell M (2011) Dacha idylls: living organically in Russia's countryside. University of California Press, Berkeley, p 40

Callaway D (1995) Resource use in rural Alaskan communities. In: Peterson DL, Johnson DR (eds) Human ecology and climate change: people and resources in the Far North. Taylor & Francis, Washington DC

Carmack E, Barber D, Christensen J, Macdonald R, Rudel B, Sakschaug E (2006) Climate variability and physical forcing of the food webs and the carbon budget on panarctic shelves. Prog Oceanogr 71(2–4):232–287

Carrington D (2017) Arctic stronghold of world's seeds flooded after permafrost melts. The Guardian, UK. Accessed 22 June 2017

Castberg R, Stokke O, Østreng W (1994) The dynamics of the Barents region. In: Stokke O, Tunander O (eds) The Barents region. Cooperation in Arctic Europe. Sage, London, pp 71–83

Castro D, Hossain K, Tytelman C (2016) Arctic ontologies: reframing the relationship between humans and Rangifer. Polar Geogr 39(2):98–112

CBD (1992) Convention on biological diversity. https://www.cbd.int/doc/legal/cbd-en.pdf. Accessed 15 July 2017

CBD (1995) Convention on biological diversity. https://www.cbd.int/decision/cop/?id=7083. Accessed 21 Jul 2017

CBD (2010) Article 8(j): traditional knowledge, innovations and practices. Convention on biological diversity. www.cbd.int/traditional/. Accessed 16 July 2017

Charles D (2006) A 'Forever' seed bank takes root in the Arctic. Science 312(5781):1730–1731

CINE (2016) Benefits of Traditional Foods." CINE. https://www.mcgill.ca/cine/research/canada/food/benefits. Accessed 10 Nov 2016

Climate Analytics (2016) What does the Paris climate agreement mean for Finland and the European Union? Technical report, June 2016. Available http://climateanalytics.org/files/ca_paris_agreement_finland_eu.pdf

Coakley JA (2003) Reflectance and albedo, surface. Encyclopedia of the atmosphere, pp 1914–1923. http://curry.eas.gatech.edu/Courses/6140/ency/Chapter9/Ency_Atmos/Reflectance_Albedo_Surface.pdf. Accessed 26 June 2017

Comas D, Reynolds R, Sajantila A (1999) Analysis of mtDNAHVRII in several human populations using an immobilized SSO probe hybridisation assay. Eur J Hum Genet 7:459–6871

Commission on Human Security (2003) "Human Security Now." New York

Coria J, Calfucura E (2012) Ecotourism and the development of indigenous communities: the good, the bad, and the ugly. Ecol Econ 73:47–55

Cózar A, Martí E, Duarte CM, García-de-Lomas J, Van Sebille E, Ballatore TJ, Eguíluz VM, González-Gordillo JI, Pedrotti ML, Echevarría F, Troublè R (2017) The Arctic ocean as a dead end for floating plastics in the North Atlantic branch of the thermohaline circulation. Sci Adv 3(4):e1600582

CRIAW (2016) Canadian Research Institute for the Advancement of Women. Defining 'Wellbeing'. Fem North Net and CRIAW ICREF. http://fnn.criaw-icref.ca/en/page/defining-wellbeing. Accessed 16 Nov 2016

CWFS (2012) Committee on world food security, "Coming to terms with terminology: food security, nutrition security, food security and nutrition, food and nutrition security." Thirty-ninth session, Rome, Italy, October 15–20, 2012, CFS 2012/39/4

Dana LP, Åge Riseth J (2011) Reindeer herders in Finland: pulled to community-based entrepreneurship & pushed to individualistic firms. Polar J 1(1):108–123

Davignon J, Gregg RE, Sing CF (1988) Apolipoprotein E polymorphism and atherosclerosis. Arteriosclerosis 8:1–21

De Pooter D (2013) Organochlorine compounds. Available from http://www.coastalwiki.org/wiki/Organochlorine_compounds. Accessed 9 May 2017

Depledge J (2000) United nations framework convention on climate change (UNFCCC) technical paper: tracing the origins of the Kyoto protocol: an article-by-article textual history (PDF), UNFCCC. http://unfccc.int/resource/docs/tp/tp0200.pdf. Accessed 12 July 2017

Dewailly E, Mulvad G, Pedersen HS, Hansen JC, Behrendt N, Hansen JPH (2003) Inuit are protected against prostate cancer. Cancer Epidemiology and Prevention Biomarkers, 12(9), 926–927

Dodds K (2013) The ilulissat declaration (2008): the Arctic states, "Law of the Sea," and Arctic ocean. SAIS Rev Int Aff 33(2):45–55

Donaldson SG, Van Oostdam J, Tikhonov C, Feeley M, Armstrong B, Ayotte P, Dallaire R (2010) Environmental contaminants and human health in the Canadian Arctic. Sci Total Environ 408(22):5165–5234

Dudarev AA, Alloyarov PR, Chupakhin VS, Dushkina EV, Sladkova YN, Dorofeyev VM, Kolesnikova TA, Fridman KB, Nilsson LM, Evengård B (2013) Food and water security issues in Russia I: food security in the general population of the Russian Arctic, Siberia and the Far East, 2000−2011. Int J Circumpolar Health 72:21848

Dudarev AA, Dushkina EV, Sladkova YN et al (2015) Evaluating health risk caused by exposure to metals in local foods and drinkable water in Pechengadistrict of Murmansk region. Meditsinatrudaipromyshlennaiaekologiia 11:25–33. (in Russian)

Duerden F (2004) Translating climate change impacts at the community level. Arctic:204–212

Duhaime G, Bernard N (2001) Regional and circumpolar conditions for food security. In: Duhaime G (ed) Sustainable food security in the Arctic: state of knowledge. CCI Press, Edmonton

Duhaime G, Chabot M, Fréchette P (1998) Portrait économique des ménages inuit du Nunaviken 1995. In: Duhaime G (ed) Les impacts socioéconomiques de la contamination de la chaîne alimentaire au Nunavik. Groupe d'études inuitet circumpolaires, Université Laval, Québec, pp 17–156

Duhaime G, Godmaire A (2001) The conditions of sustainable food security. An integrated conceptual framework. In: Duhaime G (ed) Sustainable food security in the Arctic: state of knowledge. CCI Press, Edmonton

Dulloo ME (2013) Global challenges for agricultural plant biodiversity and international collaboration. In: Conservation of tropical plant species. Springer, New York, pp 491–509

EC (2000) The European Commission. 'The European Union and the Barents region'. http://aei.pitt.edu/38747/1/A3708.pdf. Accessed 6 May 2017

EC (2016) European Commission Impact Assessment accompanying the document on the inclusion of greenhouse gas emissions and removals from land use, land use change and forestry into the 2030 climate and energy framework and amending Regulation No 525/2013 of the European Parliament and the Council on a mechanism for monitoring and reporting greenhouse gas emissions and other information relevant to climate change, Brussels

EC Canada (2012) A Climate Change Plan for the Purposes of the Kyoto Protocol Implementation Act 2012. http://www.ec.gc.ca/Publications/default.asp?lang=En&n=EE4F06AE-1&xml=EE4F06AE-13EF-453B-B633-FCB3BAECEB4F&offset=3&toc=show. Accessed 12 July 2017

EC Regulation (2002) Regulation (EC) No 178/2002 of the European Parliament and of the Council of 28 January 2002 laying down the general principles and requirements of food law, establishing the European Food Safety Authority and laying down procedures in matters of food safety. http://eur-lex.europa.eu/legal-content/en/ALL/?uri=CELEX:32002R0178. Accessed 15 July 2017

EC Russia (2017) http://ec.europa.eu/trade/policy/countries-and-regions/countries/russia/

EC-RED (2016) European Commission Renewable energy directive. https://ec.europa.eu/energy/en/topics/renewable-energy/renewable-energy-directive. Accessed 10 July 2017

Edelman M (2014) Food Sovereignty: Forgotten Genealogies and Future Regulatory Challenges, Journal of Peasant Studies 41(6): 959–978

Edin-Liljegren A, Hassler S, Sjolander P, Daerga L (2004) Risk factors for cardiovascular diseases among Swedish Saami – a controlled cohort study. Int J Circumpolar Health 63(Suppl 2):292–297

EEAS (2016) European External Action Service. Brussels, Belgium. https://eeas.europa.eu/headquarters/headquarters-homepage_en. Accessed 21 Aug 2017

EFSA (2016) European food safety authority. https://www.efsa.europa.eu/en/aboutefsa. Accessed 15 Nov 2016

Egeland GM, Harrison G (2013) "Health disparities: promoting Indigenous Peoples' health through traditional food systems and self-determination". In Indigenous peoples' food systems and well-being: interventions and policies for healthy communities, edited by H.V. Kuhnlein, B. Erasmus, D. Spigelski and B. Burlingame

Egelund GM et al (2010) Can Med Assoc J 182(3):243–248. doi:https://doi.org/10.1503/cmaj.091297

Egelund GM et al (2013) Health disparities: promoting indigenous peoples' health through traditional food systems and self-determination. In: Kuhnlein HV, Erasmus B, Spigelski D, Burlingame B (eds) Indigenous peoples' food systems and well-being, Interventions and policies for healthy communities. Food and Agriculture Organization of the United Nations, Rome, pp 9–22

Eichner JE, Dunn ST, Perveen G, Thompson DM, Stewart KE, Stroehla BC (2002) Apolipoprotein E polymorphism and cardiovascular disease: a review. Am J Epidemiol 155(6):487–495

Eilu P (2012) Mineral deposits and metallogeny of Fennoscandia. Geological Survey of Finland, Special Paper 53, 401 pages

Elenius L, Tjelmeland H, Lähteenmäki M, Golubev A (2015) In: Elenius L et al (eds) The Barents region: a transnational history of subarctic northern Europe. Pax Forlag As, Oslo, Norway, p 502

Ellis B, Brigham L (2009) Arctic marine shipping assessment 2009 report. Arctic Council, April, 2009

EMSA (2014) European maritime safety agency, ballast water. http://www.emsa.europa.eu/implementation-tasks/environment/ballast-water.html. Accessed 10 July 2017

Enattah NS, Sahi T, Savilahti E, Terwilliger JD, Peltonen L, Jarvela I (2002) Identification of a variant associated with adult type hypolactasia. Nat Genet 30:233–237

Engelhaupt E (2008) Do food miles matter? Environ Sci Technol 42:3482

Eshelby K (2015) Siberia's nomadic Nenets: home is where the pasture is. Independent. June 15, 2015. http://www.independent.co.uk/travel/europe/siberias-nomadic-nenets-home-is-where-the-pasture-is-10320703.html. Accessed 7 Nov 2016

Eskeland GS, Flottorp LS (2006) Climate change in the Arctic: a discussion of the impact on economic activity. In: Glomsrød S, Aslaksen I (eds) The economy of the north. Statistics Norway, Oslo, p 86. http://www.ssb.no/a/english/publikasjoner/pdf/sa84_en/kap6.pdf. Accessed 15 June 2017

ETLA (2016) The effects of sanctions on Finnish exports to Russia. Research Institute of the Finnish Economy, ETLA Brief 45, Helsinki, Finland

EU (2013) European Union, European Council, "Laying down the general principles and requirements of food law, establishing the European Food Safety Authority and laying down procedures in matters of food safety." EU Regulation No. 178/2002. "Facts." Barentsinfo.org. http://www.barentsinfo.org/Barents-region. Accessed 26 Aug 2016

EU Climate Action Tracker (2016) http://climateactiontracker.org/countries/eu.html. Accessed 14 Nov 2016

Europa (2017) Official website of the European Union. https://europa.eu/european-union/index_en. Accessed 15 July 2017

EVIRA (2016) Elintarviketurvallisuusvirasto. Finnish food safety authority. https://www.evira.fi/en/about-evira/about-us/activity/control/. Accessed 10 May 2017

Falkenberg K (2016) Sustainability Now! A European vision for sustainability. EPSC Strategic Notes

Fallon S (2012) Don't leave the Saami out in the cold: the Arctic-Barents region needs a binding treaty that recognizes its indigenous peoples' rights to self-determination and free, prior and informed consent. Law of the sea reports, vol. 3:1, 2012. "Finland." Grantham Research Institute on Climate Change and the Environment. http://www.lse.ac.uk/GranthamInstitute/legislation/countries/finland/. Accessed 14 Nov 2016

FAO (1996) Food and agricultural organization of the United Nations. Rome declaration on world food security and world food summit plan of action. World Food Summit 13–17. Rome, November, 1996. http://www.fao.org/docrep/003/w3613e/w3613e00.HTM

FAO (2005) Voluntary Guidelines to support the progressive realization of the Right to Adequate food in the context of National Food Security. FAO, Rome

FAO (2008) Food and agriculture organization of the United Nations. Climate change and food security: a framework document. Rome, 2008

FAO (2013) Indigenous Peoples' food systems and well-being: interventions and policies for health communities. http://www.fao.org/docrep/018/i3144e/i3144e.pdf. Accessed 10 July 2017

FAO (2014) In: Lebedys A, Li Y. (eds) Finance Working Paper FSFM/ACC/09Contribution of the forestry sector to national economies, 1990–2011. FAO, Rome

FAO (2016) Food and agriculture organization and United Nations human security unit, "human security & food security: hunger, food insecurity, and malnutrition." March 2016

FAO, IFAD and WFP (2014) The state of food insecurity in the world 2014. Strengthening the enabling environment for food security and nutrition. FAO, Rome

FAOLEX-Norway (2004) Act relating to food production and food safety, etc. (Food Act) in Norway (Lov om matproduksjon og mattrygghet mv. (matloven). http://www.fao.org/faolex/results/details/en/?details=LEX-FAOC066883. Accessed 15 July 2017

FAOLEX-Sweden (2005) Decree to implement provisions of the Food Act (SFS 1971:511). https://www.ecolex.org/details/legislation/food-decree-1971807-lex-faoc020873/. Accessed 15 July 2017

FCCA (2015) Finnish climate change act 2015. The climate change act (609/2015). http://www.ym.fi/en-us/The_environment/Legislation_and_instructions/Climate_change_legislation. Accessed 15 July 2017

FCES (2014) The Finnish committee for European security (STETE). Nordic Forum for Security Policy Final report 2014. http://www.widersecurity.fi/uploads/1/3/3/8/13383775/nordic_forum_2014.pdf. Accessed 9 July 2017

FCRN (2016) Food Climate Research Network, Oxford, UK. https://foodsource.org.uk/chapter/1-overview-food-system-challenges. Accessed 18 Jul 2017

FEA (2015) Finnish environmental agency. http://www.ymparisto.fi/en-US/Waters/Floods/Flood_risk_management/Flood_risk_management_planning. Accessed 14 July 2017

Ford J (2009) Vulnerability of Inuit food systems to food insecurity as a consequence of climate change: a case study from Igloolik, Nunavut. Reg Environ Chang 9:83–100

Forrest S (1997) Territoriality and State-Saami Relations. http://arcticcircle.uconn.edu/HistoryCulture/Saami/Saamisf.html. Accessed 12 June 2107

Fowler C (2008) The Svalbard global seed vault: securing the future of global agriculture (PDF). Global crop diversity trust. https://blogs.worldbank.org/files/climatechange/The%20Svalbard%20Seed%20Vault_Global%20Crop%20Diversity%20Trust%202008.pdf. Accessed 30 June 2017

FPH (2011) Frente Parlamentario contra el Hambre. Declaración del Segundo Foro del Frente Parlamentario contra el Hambre de América Latina y el Caribe. Available at http://www.fao.org/alc/file/media/fph/docs/fphregional/declaracion_ii_foro_fph.pdf)

Gellynck X, Banterle A, Kühne B, Carraresi L, Stranieri S (2012) Market orientation and marketing management of traditional food producers in the EU. Br Food J 114(4):481–499

General Comment 12 (1999) General comment No. 12. Committee on economic, social and cultural rights, twentieth session, 12th May 1999, United Nations Economic and Social Council, E/C.12/1999/5, (Art. 11)

General Comment 21 (2009) General comment No. 21, Committee on economic, social and cultural rights, forty-third session, 2–20 November 2009, United Nations Economic and Social Council, E/C.12/GC/21, at para. 13

General Comment 23 (1994) General comment No. 23 of the human rights committee. UN doc. A/49/40,107–110 (1994), para. 3.1

Glomsrød S, Mäenpää I, Lindholt L, McDonald H, Goldsmith S (2009) Arctic economies within the Arctic nations. In: Glomsrød S, Aslaksen I (eds) The economy of the north 2008. Statistics Norway, Oslo, Norway, pp 37–63

Glynn A, Lignell S, Darnerud PO, Aune M, Ankarberg EH, Bergdahl IA et al (2011) Regional differences in levels of chlorinated·and brominated pollutants in mother's milk from primiparous women in Sweden. Environ Int 37(1):71–79

Gordillo G, Jeronimo OM (2013) Food Security and Sovereignty (Base document for discussion). FAO, Rome. http://www.fao.org/3/a-ax736e.pdf. Accessed 14 July 2017

Goyer RA (1996) Results of lead research: prenatal exposure and neurological consequences. Environ Health Perspect 104(10):1050–1054

Graham T (2015) Encouraging Sustainable Food Choices: The Role of Information and Values in the Reduction of Meat Consumption. Victoria University of Wellington, New Zealand

Gray JS (2002) Biomagnification in marine systems: the perspective of an ecologist. Mar Pollut Bull 45:46–52

Greenpeace International (2010) Mining Impacts. April, 15, 2010. http://www.greenpeace.org/international/en/campaigns/climate-change/coal/Mining-impacts/. Accessed 1 Apr 2016

GRICCE (2013a) Greenhouse gas emission reduction (Presidential Decree 752). Grantham Research Institute on Climate Change and the Environment, 2013. http://www.lse.ac.uk/GranthamInstitute/law/greenhouse-gas-emission-reduction-presidential-decree-752/. Accessed 14 Nov 2016

GRICCE (2013b) Grantham Research Institute on Climate Change and the Environment. http://www.lse.ac.uk/GranthamInstitute/law/2020-climate-and-energy-package-contains-directive-200929ec-directive-200928ec-directive-200931ec-and-decision-no-4062009ec-of-the-parliament-and-the-council-see-below/. Accessed 27 June 2017

GRICCE (2016) Grantham Research Institute on Climate Change and the Environment (GRICCE). The global climate legislation study–2016 update. Available online at http://www.

lse.ac.uk/GranthamInstitute/wp-content/uploads/2016/11/The-Global-Climate-Legislation-Study_2016-update.pdf

Habeck JO (2002) How to turn a reindeer pasture into an oil well, and vice versa: transfer of land, compensation and reclamation in the Komi Republic. People and the land: pathways to reform in post-Soviet Siberia, pp. 125–147

Haetta OM (1996) The Sami: an indigenous people of the Arctic (trans: Gurholt OP). Davvi Girji, Vaasa, Finland

Håglin L (1991) Nutrient intake among Saami people today compared with an old, traditional Saami diet. Arctic Med Res 1991(Suppl 1):741–746

Håglin L (1999) The nutrient density of present-day and traditional diets and their health aspects: the Saami- and lumberjack families living in rural areas of Northern Sweden. Int J Circumpolar Health 58:30–43

Haldorsen T, Tynes T (2005) Cancer in the Saami population of North Norway, 1970–1997. Eur J Cancer Prev 14:63–68

Hansen M (2016) Heavy metals in food from the Norwegian, Finnish and Russian border region. Norwegian Institute for Air Research, Tromso, Norway. Presentation at "Seminar on Globalization and Food security in the Barents region. Rovaniemi, Finland. August, 16, 2016

Hansen BB, Isaksen K, Benestad RE, Kohler J, Pedersen ÅØ, Loe LE, Coulson SJ, Larsen JO, Varpe Ø (2014) Warmer and wetter winters: characteristics and implications of an extreme weather event in the high Arctic. Environ Res Lett 9:114021

Hansen JR, Korneev O, Mcbride MM, Stiansen JE (2016) Agreements concerning pollution. Barents Portal. February, 28, 2016. http://www.barentsportal.com/barentsportal/index.php/en/more/adopting-and-adapting-an-ecosystem-approach-to-management/108-supporting-legislation/644-agreements-concerning-pollution. Accessed 14 Nov 2016

Haraldson SRS (1962) Socio-medical conditions among the Lapps in northern most Sweden. Svenska Läkartidningen 59(40):2829–2844

Hartwig A (2013) Cadmium and cancer. In: Cadmium: from toxicity to essentiality. Springer, Dordrecht, pp 491–507

Hassler S, Sjölander P, Barnekow-Bergqvist M, Kadesjö A (2001) Cancer risk in the reindeer breeding Saami population of Sweden, 1961–1997. Eur J Epidemiol 17:969–976

Hassler S, Sjölander P, Grönberg H, Johansson R, Damber L (2008) Cancer in the Saami population of Sweden in relation to lifestyle and genetic factors. Eur J Epidemiol 23:273–280

Haugen HM (2012) International obligations and the right to food: clarifying the potentials and limitations in applying a human rights approach when facing biofuels expansion. J Hum Rights 11(3):405–429

Havas K, Adamsson K, Sievers K (2015) Hungry for Finland eat local, eat slow, eat pure, eat wild: Finland's first food tourism strategy. Haaga-Helia University of Applied Science, 2015

Heikkinen H (2006) Neo-entrepreneurship as an adaptation model of reindeer herding in Finland. Nomadic Peoples 10:187–208

Heikkinen HI, Sarkki S (2010) Global area conservation ideals versus the local realities of Reindeer Herding in Northernmost Finland. Int J Bus Global 4:2

Heimstad ES, Sandanger T (2013) Food and health security in the Norwegian, Finnish and Russian border region: linking local industries, communities and socio-economic impacts. Presentation at the Barents health conference, Kirkenes, Norway, 12 September 2013. http://kolarctic.nilu.no/wp-content/uploads/2013/07/KO467-Food-and-health-security-100913-ESH.pdf. Accessed 1 July 2017

Heininen L, Nicol HN (2007) A new northern security agenda. In: Brunet-Jailly E (ed) Borderlands: comparing border security in North America and Europe. University of Ottawa Press, Canada

Heleniak T (1999) Out-migration and depopulation of the Russian north during the 1990s. Post Sov Geogr Econ 40:155–205

Henson SA, Beaulieu C, Ilyina T, John JG, Long M, Séférian R, Tjiputra J, Sarmiento JL (2017) Rapid emergence of climate change in environmental drivers of marine ecosystems. Nat Commun 8:14682

Hermeling C, Klement JH, Koesler S, Köhler J, Klement D (2015) Sailing into a dilemma: an economic and legal analysis of an EU trading scheme for maritime emissions. Transp Res A Policy Pract 78:34–53

Herrmann TM, Sandström P, Granqvist K, D'Astous N, Vannar J, Asselin H, Saganash N, Mameamskum J, Guanish G, Loon JB, Cuciurean R (2014) Effects of mining on reindeer/caribou populations and indigenous livelihoods: community-based monitoring by Sami reindeer herders in Sweden and First Nations in Canada. Polar J 4(1):28–51

Himanen S et al (2012) Analysis of regional climate strategies in the Barents region. Reports of the Ministry of the Environment, 2012. "Historical Roots of the North." Barentsinfo.org. http://www.barentsinfo.org/Contents/History. Accessed 7 Nov 2016

Hoel AH (2009) Best Practices in ecosystem-based ocean management in the Arctic. Norsk Polarinstitutt. Norwegian Polar Institute, Polar Environmental Centre, Tromsø, Norway. https://rafhladan.is/bitstream/handle/10802/8652/media.pdf?sequence=2. Accessed 30 Jul 2017

Hof AR, Roland J, Nilsson C (2015) Future of biodiversity in the Barents region. Vol. 2015519. Nordic Council of Ministers, 2015

Holt-Giménez E, Shattuck A (2011) Food crises, food regimes and food movements: rumblings of reform or tides of transformation? J Peasant Stud 38(1):109–144

Hossain K (2015a) Invasive species in the Arctic-Barents: concerns, regulation and governance. In: Pincus R, Ali SH (eds) Diplomacy on ice: energy and the environment in the Arctic and Barents. Yale University Press, New Haven

Hossain K (2015b) Cultural rights as collective rights an international law perspective. In: Jakubowski A (ed) Studies in intercultural human rights, vol 7. Brill Nijhoff, Leiden, pp 113–132

Hossain K (2015c) Invasive species in the Arctic: concerns, regulations, and governance. In: Pincus R, Ali SH (eds) Diplomacy on ice: energy and the environment in the Arctic and Antarctic. Yale University Press, New Haven. http://www.jstor.org/stable/j.ctt13x1stj

Hossain K, Koivurova T, Zojer G (2014) Understanding risks associated with offshore hydrocarbon development. In: Tedson E, Cavalieri S, Andreas Kraemer R (eds) Arctic marine governance: opportunities for transatlantic cooperation. Springer, Berlin

Hovelsrud GK et al (2012) Arctic societies, cultures, and peoples in a changing cryosphere. Ambio 40(Suppl 1):100–110

Hueffer K, Parkinson AJ, Gerlach R, Berner J (2013) Zoonotic infections in Alaska: disease prevalence, potential impact of climate change and recommended actions for earlier disease detection, research, prevention and control. Int J Circumpolar Health 72. https://doi.org/10.3402/ijch.v72i0.19562

Hukkinen J, Müller-Wille L, Heikkinen H (2003) Development of participatory institutions for reindeer management in northern Finland: preliminary synthesis and report. Helsinki University of Technology, Espoo, Finland

Huntington H, Kankaanpää P, Baldursson S, Sippola AL, Kaitala S, Zöckler C (2001) Arctic flora and fauna: status and conservation. Conservation of Arctic Flora and Fauna, 2001

IAEA (2003) International atomic energy agency, "Modelling of the radiological impact of radioactive waste dumping in the Arctic Seas." Report of the modelling and assessment working group of the international Arctic seas assessment project (IASAP), January 2003

IASC (2008) International Arctic science committee's report to SAOs, submitted at the Arctic council senior Arctic officials meeting in Kautokeino, Norway, 19–20 November 2008

ICC (2012) Food security across the Arctic. Background paper of the steering committee of the circumpolar Inuit health strategy. Inuit Circumpolar Council, Canada. http://www.inuitcircumpolar.com/uploads/3/0/5/4/30542564/icc_food_security_across_the_arctic_may_2012.pdf. Accessed 26 May 2017

ICR (2017) The international centre for reindeer husbandry. http://reindeerherding.org/herders/sami-norway/. Accessed 23 May 2017

ILO C169 (1989) Articles 2, 3, 6, 7 of the international labour organization - indigenous and tribal peoples convention, (No 169), Geneva, international labour organization (ILO), indigenous and tribal peoples convention, C169, 27 June 1989, C169. http://www.ilo.org/wcmsp5/groups/public/@ed_norm/@normes/documents/publication/wcms_100792.pdf

Incopera FP (2016) Climate change: a wicked problem. Complexity and uncertainty at the inter-section of science, economics, politics, and human behavior. Cambridge University Press, New York, 370 p

Ingvaldsen R, Loeng H, Asplin L (2002) Variability in the Atlantic inflow to the Barents Sea based on a one-year time series from moored current meters. Cont Shelf Res 22(3):505–519

Interview with Bruce Forbes. Rovaniemi, Finland, 24 Nov 2016

Interview with Juha Joona, Rovaniemi, Finland, 10 Nov 2016a

Interview with Juha Joona. Rovaniemi, Finland, 29 Nov 2016b

IPCC (2013) In: Stocker TF, Qin D, Plattner G-K, Tignor M, Allen SK, Boschung J, Nauels A, Xia Y, Bex V, Midgley PM (eds) Climate change 2013: the physical science basis. Contribution of working group I to the fifth assessment report of the intergovernmental panel on climate change. Cambridge University Press, Cambridge, United Kingdom and New York, NY, USA, p 1535

Itkonen TI (1948) Suomen lappalaiset vuoteen 1945 (Finland's Saami to the year 1945), vol 1–2. Werner Soderstrom, Porvoo/Helsinki

Jahr EH (1996) On the pidgin status of Russenorsk. In: Jahr EH, Broch I (eds) Language contact in the Arctic: northern pidgins and contact languages. Mouton de Gruyter, Berlin, pp 107–122

Jansson R et al (2015) Future changes in the supply of goods and services from natural ecosys-tems: prospects for the European north. Ecol Soc 20(3):32

Jarratt E (2014) High north food crisis, 20 October 2014. http://barentsobserver.com/en/nature/2014/10/high-north-food-crisis-20-10. Accessed 10 Nov 2016

Jordana J (2000) Traditional foods: challenges facing the European food industry. Food Res Int 33:147–152

Kelly J, Lanier A, Santos M, Healey S, Louchini R, Friborg J, Kon Y (2008) Cancer among the circumpolar Inuit, 1989–2003. II. Patterns and trends. Int J Circumpolar Health 67(5):408–420

Kelman I, Næss MW (2013) Climate change and displacement for indigenous communities in Arctic Scandinavia. The Brookings-LSE Project on Internal Displacement. Washington DC

Kettle DF (ed) (2002) Environmental governance. A Report on the next generation of environmen-tal policy. Brookings Institution, Washington

Kim C, Chanf HM, Receveur O (1998) Risk assessment of cadmium exposure in fort resolution, Northwest Territories, Canada. Food Addit Contam 15(3):307–317

Kirby A et al (2013) Mercury–Time to act United Nations Environment Programme, 2013

Kireeva A (2016) Norwegian politicians and citizens call Norilsk Nickel 'dirtiest industry in the Arctic-Barents'. August, 11, 2016. http://bellona.org/news/industrial-pollution/the-kola-min-ing-and-metallurgy-combine/2016-08-norwegian-politicians-and-citizens-call-norilsk-nickel-dirtiest-industry-in-the-Arctic-Barents. Accessed 9 Nov 2016

Kiviranta H IV, Hallikainen A, Ovaskainen ML, Kumpulainen J, Vartiainen T (2001) Dietary intakes of polychlorinated dibenzo-p-dioxins, dibenzofurans and polychlorinated biphenyls in Finland. Food Addit Contam 18(11):945–953

Klein A (2016) Ocean cleaning sea bins will gobble up plastic waste to recycle. New Scientist, 29 July 2016. https://www.newscientist.com/article/2099339-ocean-cleaning-sea-bins-will-gobble-up-plastic-waste-to-recycle/. Accessed 28 June 2017

Knuth L (2009) The right to adequate food and indigenous peoples: how can the right to food benefit indigenous peoples? Food and Agriculture Organization of the United Nations, Rome

Kohllechner-Autto M (2011) Strategic tourism development in the Barents region – an analysis. Public – private partnership in Barents tourism. Lapland Institute for Tourism Research and Education.   http://matkailu.luc.fi/loader.aspx?id=bc226714-a356-4997-ad30-94ca1f1593f2. Accessed 12 May 2017

Koivurova T, VanderZwaag D (2007) The Arctic council at 10 years: retrospect and prospects. Univ B C Law Rev 40(1):121–194. Available at SSRN: https://ssrn.com/abstract=1860308. Accessed 2 June 2017

Komi. BarentsInfo.org. http://www.barentsinfo.org/Contents/Indigenous-people/Komi. Accessed 6 Nov 2016

Kozlov A, Borinskaya S, Vershubsky G, Vasilyev E, Popov V, Sokolova M et al (2008) Genes related to the metabolism of nutrients in the Kola Sami population. Int J Circumpolar Health 67(1):58–68

Kozlov A, Lisitsyn D (1997) Hypolactasia in Saami subpopulations of Russia and Finland. Anthropol Anz 55:281–287

Krauss C et al (2005) As polar ice turns to water, dreams of treasure abound. The New York Times, October 10, 2005

Kuhnlein HV, Chan HM (2000) Environment and contaminants in traditional food systems of northern indigenous peoples. Ann Rev Nutr 20(1):595–626

Kuhnlein HV, Receveur O (1996) Dietary change and traditional food systems of indigenous peoples. Ann Rev Nutr 16(1):417–442

Kuhnlein HV, Burlingame B (2013) Why do Indigenous Peoples' food and nutrition interventions for healthy promotion and policy need special consideration? In: Kuhnlein HV et al (eds) Indigenous peoples' food systems & well-being: interventions & policies for healthy communities. FAO, Rome

Kuhnlein HV, Chan HM (2000) Environment and contaminants in traditional food systems of northern indigenous peoples. Annu Rev 20:595–626

Kullerud L, Ræstad N (2016) Oil and gas resources in the Barents sea. Grid Arendal. http://www.grida.no/publications/et/at/page/2543.aspx. Accessed 9 Nov 2016

La Rue FW, Elham M (2015) I am very active in the Internet, in order to sensitize people on our theme: freedom of expression, access to public information, right to education, freedom of association, development, pp. 82–83

Laird SA (2002) In: Laird SA (ed) Biodiversity and traditional knowledge: equitable partnerships in practice. Earthscan, London, p 504

Lang T (2017) Re-fashioning food systems with sustainable diet guidelines: towards a SDG 2 strategy), Food Climate Research Network, Oxford, UK, April 2017

Länsman et al (1992) Länsman et al v Finland, Communication No. 511/1992. https://www.escr-net.org/caselaw/2006/lansman-et-al-v-finland-communication-no-5111992-un-gaor-52nd-session-un-doc-ccpr-c-52d. Accessed 16 Aug 2017

Lansman AS (1999) Suomalaisten erii matkailijoiden ja saamelairten kohtaaminen -pullon henki vuorovaikutuksessa (The Meeting of Finnish nature travellers and Sàmi: the bottle's spirit in interaction): PhD thesis proposal, unpublished manusctipt. Anâr

Lassuy DR and Lewis PN (2013) Invasive species: human-induced. Arctic biodiversity assessment. Status and trends in arctic biodiversity, pp 559–565

Lawrence R, Raitio K (2006) Forestry conflicts in Finnish Sapmi: local, national and global links. Indigenous affairs, 2006

Lehtinen S, Luoma P, Nayha S, Hassi J, Ehnholm C, Nikkari T et al (1998) Apolipoprotein A-IV polymorphism in Saami and Finns: frequency and effect on serum lipid levels. Ann Med 30:218–223

Liotta PH, Owen T (2006) Why human security? Whitehead J Diplomacy Int Relat VII(1):37–55

Liu N, Hossain K (2017) In: Koivurova T, Qin T, Nykänen T, Dyuck S (eds) Arctic law and governance: the role of China and Finland. Hart Publishing Limited, Oxford, pp 233–251

London Convention (1972) Convention on the prevention of marine pollution by dumping of waste and other matter "London Convention" http://www.imo.org/en/OurWork/Environment/LCLP/Documents/LC1972.pdf. Accessed 27 July 2017

London Protocol (1996) Protocol to the convention on the prevention of marine pollution by dumping of wastes and other matter. http://www.imo.org/en/OurWork/Environment/LCLP/Pages/default.aspx). Accessed 27 July 2017

Lougheed T (2010) The changing landscape of Arctic traditional food. Environ Health Perspect 118(9):A386–A393

Lundqvist J, Grönwall J, Jägerskog A (2015) Water, food security and human dignity – a nutrition perspective. Ministry of Enterprise and Innovation, Swedish FAO Committee, Stockholm, 2015

Macarthur E (2016) More plastic than fish in the sea by 2050. The new plastics economy: rethinking the future of plastics. https://www.ellenmacarthurfoundation.org/publications/the-new-plastics-economy-rethinking-the-future-of-plastics. Accessed 26 Aug 2017

Mace PM, Gabriel WL (1999) Evolution, scope, and current applications of the precautionary approach in fisheries, proceedings, 5th NMFS NSAW. 1999. NOAA Tech. Memo. NMFS-F/SPO-40, at 65, http://www.st.nmfs.noaa.gov/Stock Assessment/workshop_documents/nsaw5/mace_gab.pdf. Accessed 06 July 2017

Magga OH (2006) Diversity in Saami terminology for reindeer, snow, and ice. Int Soc Sci J 58(187):25–34

Magni G (2016) Paper commissioned for the global education monitoring, report, education for people and planet: creating sustainable futures for all. UNESCO

Mahoney MC, Michalek AM (1991) A meta-analysis of cancer incidence in United States and Canadian native populations. Int J Epidemiol 20:323–237

Malthus TF (1826) An essay on the principle of population, 2. John Murray, London. http://oll.libertyfund.org/titles/malthus-an-essay-on-the-principle-of-population-vol-1-1826-6th-ed. Accessed 17 July 2017

MARPOL (2016) International Convention for the Prevention of Pollution form ships (MARPOL). https://www.lr.org/en/marpol-international-convention-for-the-prevention-of-pollution/. Accessed 26 Jul 2017

Mason A (2000) Community, solidarity and belonging: levels of community and their normative significance. Cambridge University Press, Cambridge

McDonald-Gibson C (2013) The shipping forecast – it'll be colder but much, much quicker: new Arctic-Barents shipping route saves up to two weeks' travel between Asia and Europe. The independent, September 11, 2013, http://www.independent.co.uk/news/world/europe/the-shipping-forecast-it-ll-be-colder-but-much-much-quickernew-Arctic-Barents-shipping-route-saves-up-to-8810085.html. Accessed 1 April 2016

McRae R, Hubert D (2001) Human Security and New Diplomacy. McGill-Queen's University Press, Montreal

Meakin S, Kurvits T (2009) In: Affairs IN (ed) Assessing the impacts of climate change on food security in the Canadian Arctic. GRID-Arendal, Ottawa, Canada, p 46

Miettinen S (2006) Raising the status of Lappish communities through tourism development. In: Cultural tourism in a changing world: politics, participation and (re) presentation. Ed. by Melanie K. Smith and Mike Robinson Channel View Publications Bristol,

Minamata Convention (2017) Minamata convention on mercury. http://www.mercuryconvention.org/Convention. Accessed 18 July 2017

MMM (2016) Maa ja metsätalousministeriö. Finnish ministry of agriculture and forestry. http://mmm.fi/etusivu. Accessed 17 Aug 2017

Moe A (2010) Russian and Norwegian petroleum strategies in the Barents Sea. Arctic Rev 1(2):225–248

Muir DC, Wagemann R, Hargrave BT, Thomas DJ, Peakall DB, Norstrom RJ (1992) Arctic marine ecosystem contamination. Sci Total Environ 122(1-2):75–134

Müller D, Petterson R (2001) Access to Sami tourism in northern Sweden. Scand J Hosp Tour 1(1):5–18

Muller A, Bautze L, Meier M, Gottinger A (2016) Organic farming, climate change, mitigation and beyond. Reducing the environmental impacts of EU agriculture. IFOAM EU Group, Brussels, Belgium

Muller-Wille L (2001) From reindeer stew to pizza: the displacement of local food resources in Sampi, Northernmost Europe. In: Sustainable food security in the Arctic-Barents: state of knowledge. Ed. Gerard Duhaime. CCI Press Edmonton 2001

Muller-Wille L et al (2008) Community viability and well-being in northernmost Europe: social change and cultural encounters, sustainable development and food security in Finland's North. Int J Bus Global 2:4

Murphy NJ, Schraer CD, Thiele MC, Boyko EJ, Bulkow LR, Doty BJ et al (1995) Dietary change and obesity associated with glucose intolerance in Alaska Natives. J Am Diet Assoc 95(6):676–682

Mustonen T, Jones G (2015) Reindeer herding in Finland

Nachmany M, Fankhauser S, Davidová J, Kingsmill N, Landesman T, Roppongi H, Schleifer P, Setzer J, Sharman A, Singleton CS, Sundaresan J (2015a) Sweden: the 2015 global climate legislation study. A review of climate change legislation in 99 countries

Nachmany M, Fankhauser S, Davidová J, Kingsmill N, Landesman T, Roppongi H, Schleifer P, Setzer J, Sharman A, Singleton S, Sundaresan J, Townshend T (2015b). "Russia." The 2015 global climate legislation study: a review of climate change legislation in 99 countries, 2015

NEA (2015) Norwegian Environment Agency. http://www.miljodirektoratet.no/Documents/publikasjoner/M418/M418.pdf. Accessed 12 July 2017

Neuvonen S, Viiri H (2017) Changing climate and outbreaks of forest pest insects in a cold Northern Country, Finland. In: Latola K, Savela H (eds) The Interconnected Arctic — UArctic Congress 2016. Springer, Cham

NFA (2015) National food agency, sweden. https://www.livsmedelsverket.se/en/about-us/. Accessed 2 May 2017

NFSA (2016) Norwegian food safety authority. https://www.mattilsynet.no/language/english/about_us/. Accessed 2 May 2017

NFSP (2014) The Finnish committee for European Security, "The Arctic and Barents Regions: cooperation, human rights and security challenges." Nordic forum for security policy, final report, 2014

Nilsen T (2010) Multi-billions investments in the Barents region. November, 24, 2010. http://barentsobserver.com/en/sections/russia/multi-billion-investments-barents-region. Accessed 24 Nov 2010

Nilsen T (2016), Airborne pollution exceeds federal limits by 10 times. The Barents Observer, 12 July 2016. http://thebarentsobserver.com/ecology/2016/07/airborne-pollution-exceeds-federal-limits-10-times. Accessed 9 Nov 2016

Nilsen H, Utsi E, Bønaa KH (1999) Dietary and nutrient intake of a Saami population living in traditional Reindeer herding areas in north Norway: comparisons with a group of Norwegians. Int J Circumpolar Health 58:120–133

Nilsson LM (2012) Saami lifestyle and health: epidemiological studies from northern Sweden. Umeå University, 2012

Nilsson AE, Carlsen H and van der Watt LM (2015) Uncertain futures: the changing global context of the European Arctic, Report of a Scenario-building Workshop in Pajala, Sweden. Working Paper no. 2015-12. Stockholm Environment Institute, Stockholm

Nilsson LM, Evengård B (2013) Food and water security indicators in an Arctic health context. Umeå University, 2013

Nilsson LM, Evengård B (2015) Food security or food sovereignty: what is the main issue in the Arctic? in the new Arctic. Springer International Publishing, Berlin, pp 213–223

Nilsson LM et al (2011) Diet and lifestyles of the Saami of southern Lapland in the 1930s –1950s and today. Int J Circumpolar Health 70(3):301–318

Noble B, Hanna K (2015) Environmental assessment in the Arctic: A gap analysis and research agenda. ARCTIC 68(3):341–55

Nordstrom K et al (2013) Food and health: individual, cultural, or scientific matters? Genes Nutr 8:357–363. "Norway." Grantham Research Institute on Climate Change and the Environment. http://www.lse.ac.uk/GranthamInstitute/legislation/countries/norway/. Accessed 14 Nov 2016

Norstrom RJ, Belikov SE, Born EW, Garne GW, Malone B, Olpinski S, Ramsay MA et al (1998) Chlorinated hydrocarbon contaminants in polar bears from eastern Russia, North America, Greenland and Svalbard: biomonitoring of Arctic pollution. Arch Environ Contam Toxicol 35:354–367

Northern Dimension (2016) About ND Northern Dimension. http://www.northerndimension.info/northern-dimension. Accessed 14 Nov 2016

Nøst TH, Breivik K, Fuskevåg OM, Nieboer E, Odland JØ, Sandanger TM (2013) Persistent organic pollutants in Norwegian men from 1979 to 2007: intraindividual changes, age–period–cohort effects, and model predictions. Environ Health Perspect 121(11-12):1292

Nowlan L (2001) Arctic legal regime for environmental protection. No. 44. IUCN, 2001

Nuttall M (1992) Arctic Homeland: kinship, community and development in Northwest Greenland. University of Toronto Press, Toronto

Nuttall M (2000a) The Arctic is changing. The Arctic – a web resource on human-environment relationships in the Arctic. http://www.thearctic.is/PDF/The%20Arctic%20is%20changing.pdf. Accessed 12 June 2017

Nuttall M (2000b) The Arctic is changing. Stefansson Arctic institute and individual authors

Nuttall M (2012) Encyclopedia of the Arctic, vol 1–3. Routledge, New York

Nuttall, M., Berkes, F., Forbes, B., Kofinas, G., Vlassova, T., Wenzel, G. 2005. "Hunting, herding, fishing and gathering: indigenous peoples and renewable resource use in the Arctic". In Arctic Climate Impact Assessment. Cambridge University Press

Nyborg M (2013) Norwegian White Paper Climate Change Adaptation. http://arcticadaptationexchange.com/share/norwegian-white-paper-climate-change-adaptation-0. Accessed 13 July 2017

Nyéléni Newsletter (2013) Nyéléni Declaration on Food Sovereignty, no 13, March 2013. www.nyeleni.org. Accessed 15 July 2017

Nyeleni (2016) Nyéléni Pan-European Forum Report.European Food Sovereignty Movement. http://www.nyelenieurope.net/publications. Accessed 10 Apr 2017

Nystén-Haarala S and Kulysasova A (2012) Rights to traditional use of resources in conflict with legislation: a case study of pomor fishing villages on the white sea coast. Politics of development in the Barents region; Tennberg, M. (ed) Lapland University Press: Rovaniemi, Finland, pp 1–386. Available http://www.doria.fi/bitstream/handle/10024/86168/Politics_of_development_doria.pdf?sequence=1. Accessed 13 Jul 2017

O'Loughlin J, Kolossov V, Radvanyi J (2007) The Caucasus in a time of conflict, demographic transition, and economic change. Eurasian Geogr Econ 48(2):135–156

Odland JØ, Deutch B, Hansen JC, Burkow IC (2003) The importance of diet on exposure to and effects of persistent organic pollutants on human health in the Arctic. Acta Paediatr 92(11):1255–1266

OECD (2016a) Agricultural policy monitoring and evaluation. OECD Publishing, Paris. https://doi.org/10.1787/agr_pol-2016-en. Accessed 13 July 2017

OECD (2016b) Organisation for economic co-operation and development ministerial meeting, April 2016. https://www.oecd.org/tad/events/COP21-paris-agreement-and-agriculture-draft.pdf. Accessed 14 July 20178

OHCHR (2013) United nations security council resolution 66/290 (2013). United Nations human rights office of the high commissioner, "What are human rights?" http://www.ohchr.org/EN/Issues/Pages/WhatareHumanRights.aspx. Accessed 7 Nov 2016

OHCHR (2016) Office of the United Nations high commissioner for human rights, "The right to adequate food." Fact sheet No. 34. Office of the United Nations high commissioner for human rights, "Toolkit on the right to food." http://www.ohchr.org/EN/Issues/ESCR/Pages/Food.aspx. Accessed 25 April 2016

One Health Initiative (2016) "One health". Centres for Disease Control and Prevention (CDC), USA. https://www.cdc.gov/onehealth/index.html. Accessed 3 Jan 2017

OPM (2014) Cultural cooperation in the Barents region. Strategy 2014–2018. Opetus- ja kulttuuriministeriö. Publications of the Ministry of Education and Culture, p. 7

Ortega H, Castilla P, Gomez-Coronado D, Garces C, Benavente M, Rodriguez-Artalejo F et al (2005) Influence of apolipoprotein E genotype on fat-soluble plasma antioxidants in Spanish children. Am J Clin Nutr 81:624–632

Oslo Protocol (1994) Protocol on Further Reduction of Sulphur Emissions. https://www.unece.org/env/lrtap/fsulf_h1.html. Accessed 27 July 2017

OSPAR Commission (2000) Quality status report 2000, OSPAR Commission, London

Ottawa Declaration (1996) Declaration on the establishment of the Arctic Council. https://oaarchive.arctic-council.org/handle/11374/85. Accessed 23 July 2017

Owen T (2004) Challenges and opportunities for defining and measuring human security, Disarmament Forum 3(June 2004):17

Paci C, Dickson C, Nickels S, Chan L, Furgal C (2004) Food security of northern indigenous peoples in a time of uncertainty. In: 3rd northern research forum open meeting, Yellowknife, 2004

Palmer L (2013) Melting Arctic ice will make way for more Ships and more species invasions. Scientific American, March 6, 2013, http://www.scientificamerican.com/article/melting-Arctic-sea-ice-means-more-shipping-and-moreinvasive-species/. Accessed 1 Apr 2016

PAME (2013) The Arctic ocean review project, final report, (Phase II 2011–2013), Kiruna May 2013. Protection of the Arctic marine environment (PAME) Secretariat, Akureyri (2013) https://cil.nus.edu.sg/wp/wpcontent/uploads/2014/06/2013_Arctic_Ocean_Review_Phase_II_Report.pdf. Accessed 15 July 2017

Pavlids C, Patrinos GP, Katsila T (2015) Nutrigenomics: a controversy. Appl Transl Genom 4:50–53

Perry MJ (2005) The morality of human rights: a non-religious ground. Dublin ULJ 27:28

Pettersen O (2002) The vision that became reality: the regional Barents cooperation, 1993–2003: the Barents Euro-Arctic Region. Barents Secretariat

Picq M (2012) Listening to the Arctic. Aljazeera, September, 27, 2012. http://www.aljazeera.com/indepth/opinion/2012/09/2012926105424921519.html. Accessed 2 Apr 2016

Pimbert M (2009) Towards food sovereignty. International Institute for Environment and Development, London

Pinstrup-Andersen P (2009) Food security: definition and measurement. Food Secur 1(1):5–7

Plummer R, Baird J (2013) Adaptive co-management for climate change adaptation: considerations for the Barents region. Sustainability 5:629–642

Polar Code (2014) International code for ships operating in polar waters (Polar Code). http://www.imo.org/en/MediaCentre/HotTopics/polar/Documents/POLAR%20CODE%20TEXT%20AS%20ADOPTED%20BY%20MSC%20AND%20MEPC.pdf. Accessed 27 July 2017

Porta L, Abou-Abssi E, Dawson J, Mussells O (2017) Shipping corridors as a framework for advancing marine law and policy in the Canadian Arctic. Ocean Coast Law J 22:63

Puglielli L, Tanzi RE, Kovacs DM (2003) Alzheimer's disease: the cholesterol connection. Nat Neurosci 6(4):345–351. https://doi.org/10.1038/nn0403-345. PMID 12658281. Accessed 24 September 2017

Pursiainen C (2001) Soft security problems in Northwest Russia and their implications for the outside world. A framework for analysis and action, 31

Rafaelsen R (2010) The Barents cooperation. New regional approach to foreign policy in the high north – in Barents review 2010. In: Staalesen A (ed) Talking Barents. People, borders and regional cooperation. The Norwegian Barents Secretariat, Kirkenes, pp 25–31

Raheem D (2016) Traditional foods and their relevance to food security in the Barents region. "Globalization and Food security in the Barents region" seminar, Arctic Centre/NIEM, University of Lapland, Rovaniemi, Finland. August 16, 2016

Rahman NA, Saharuddin AH, Rasdi R (2014) Effect of the northern sea route opening to the shipping activities at Malacca straits. Int J e-Navig Marit Econ 1:85–98

Raskin P, Banuri T, Gallopin G, Gutman P, Hammond A (2002) Great transition: the promise and lure of the times ahead. Stockholm Environment Institute, Stockholm

Rautio R, Bambuyalk A, Hahl M (2013) Economic cooperation in the Barents region. AkavaplanNiva AS report no: 6265. http://www.barentsinfo.org/BC/Barentskeskus/Economy2013. Accessed 5 May 2017

Rautio A et al (2017) Healthy living, nutrition and food waste in the Barents region. NOFIMA Report 5/2017. https://brage.bibsys.no/xmlui/bitstream/handle/11250/2435199/Rapport%2B05-2017.pdf?sequence=1&isAllowed=y. Accessed 21 May 2017

Rees WG, Stammler FM, Danks FS, Vitebsky P (2008) Vulnerability of European reindeer husbandry to global change. Clim Chang 87:199–217. https://doi.org/10.1007/s10584-007-9345-1. Accessed 19 September 2017

Reist JD, Power M, Dempson JB (2013) Arctic charr (*Salvelinus alpinus*): a case study of the importance of understanding biodiversity and taxonomic issues in northern fishes. Biodiversity 14:45–56

Renko S, Renko N, Polonijo T (2010) Understanding the role of food in rural tourism development in a recovering economy. J Food Prod Mark 16:309–324

Revich B, Tokarevich B, Arkinson AJ (2012) Climate change and zoonotic infections in the Russian Arctic. Int J Circumpolar Health 71:18792. https://doi.org/10.3402/ijch.v71i0.18792

Revolvy (2017) Sami people. https://www.revolvy.com/topic/Sami+People. Accessed 16 June 2017

RFSSS (2011) Russian federal state statistics service. "Всероссийская перепись населения 2010 года. Том 1" [2010 All-Russian Population Census, vol. 1]. Всероссийская перепись населения 2010 года (2010 All-Russia population census) (in Russian). Federal state statistics service. http://www.gks.ru/. Accessed 27 June 2017

Rheen S (1983) En kortt Relation om Lapparnes Lefwarne och Sedher, wijd-Skiepellsser, sampt i många Stycken Grofwe wildfarellsser. In: Wiklund KB (ed) Berättelser om samerna i 1600-talets Sverige. Umeå: Kungl. Skytteanska samfundet. Originally published in Bidrag till kännedom om de svenska landsmålen ock svenskt folkliv, XVII. 1. Harald Wretmans tryckeri, Uppsala, pp 7–68

Robards M (2013) Resilience of international policies to changing social-ecological systems: Arctic shipping in the Bering Strait. In: Arctic resilience Interm report 2013, Stockholm Environment Institute and the Stockholm Resilience Centre

Roberts J (2007) Marine environment protection and biodiversity conservation: the application and future development of the IMO's particularly sensitive sea area concept 23 (2007)

Rockström J, Gaffney O, Rogelj J, Meinshausen M, Nakicenovic N, Schellnhuber HJ (2017) A roadmap for rapid decarbonization. Science 24, 355(6331). https://doi.org/10.1126/science.aah3443

Rohe R (1986) The evolution of the great lakes logging camp, 1830-1930. J For Hist 30(1):17–28

Rohr J (2014) Indigenous peoples in the Russian federation. Report 18 IWGIA, 2014

Romild U, Fredman P, Wolf-Watz D (2011) Socio-economic determinants, demand and constraints to outdoor recreation participation in Sweden. Friluftsliv i förändring, Ostersund, Sweden

Ross AB (2009) Adherence to a traditional lifestyle affect food and nutrient intake among modern Swedish Saami. Int J Circumpolar Health 68(4):2009

Ross AB, Johansson A, Ingman M, Gyllensten U (2006) Lifestyle, genetics, and disease in Saami. Croat Med J 47(4):553–565

Rothwell DR (2000) Global environmental protection instruments. In: Vidas Davor, Protecting the polar marine environment: law and policy for pollution prevention, Cambridge University Press, Cambridge. , "Russian Federation." Climate Action Tracker. November 3, 2016. http://climateactiontracker.org/countries/russianfederation.html. Accessed 14 Nov 2016

Rottem SV, Moe A (2007) Climate change in the north and the oil industry. Input to strategic impact assessment. Barents region, 2030

Rover C, Ridder-Strolis K (2014) A sustainable Arctic: preconditions, pitfall and potentials. Länderbericht. http://www.kas.de/wf/doc/kas_39168-1522-2-30.pdf?161026150640. Accessed 4 June 2017

Ruong I (1969) "Samema" (The Saami). Aldusserien 268. Stockholm: Bokf6rlaget/Bonniers

Russell BA (2008) The Arctic environmental protection strategy & the new Arctic council. Archived from the original on 11 October 2008. http://arcticcircle.uconn.edu/NatResources/Policy/uspolicy1.html. Accessed 23 May 2017

Russian Federation (2016) The climate action tracker. http://climateactiontracker.org/countries/russianfederation.html. Accessed 21 July 2017

Saami Y (2017) Activists protest against Teno river fishing rules. https://yle.fi/uutiset/osasto/news/sami_activists_occupy_island_in_protest_at_tenojoki_fishing_rules/9717663. Accessed 30 Jul 2017

Sahi T (1994) Genetics and epidemiology of adult-type hypolactasia. Scand J Gastroenterol Suppl 202:7–20. "Sámi – Norway." Reindeer herding: a virtual guide to reindeer and the people who herd them. http://reindeerherding.org/herders/Saami-norway/. Accessed 10 Nov 2016

Sara MN (2009) Siida and traditional Sami reindeer herding knowledge. (General Articles) (Report). Northern Review. https://www.highbeam.com/doc/1G1-202252650.html. Accessed 23 May 2017

Schanbacher WD (2010) The politics of food: the global conflict between food security and food sovereignty. ABC-CLIO, Santa Barbara

Schauer U, Loeng H, Rudels B, Ozhigin VK, Dieck W (2002) Atlantic water flow through the Barents and Kara Seas. Deep Sea Res Part I Oceanogr Res Pap 49:2281–2298

Sellheim N (2011) The Barents environmental cooperation: a legitimacy analysis. University of Akureyri, April 2011

Sellheim N (2017) The reflection of multilateral environmental agreements (MEAs) in the Barents environmental cooperation. Arctic Rev 3(2.) ISSN 2387-4562. Available https://arcticreview. no/index.php/arctic/article/view/37. Accessed 30 July 2017

Sen A (1989) Food and freedom. World Develop 17(6):769–781. Elsevier, UK

SEPA (2016) Swedish environmental protection agency. http://www.swedishepa.se/Guidance/ Laws-and-regulations/The-Swedish-Environmental-Code/. Accessed 14 July 2017

Serreze MC, Walsh JE, Chapin FS III, Osterkamp T, Dyurgerov M, Romonovsky V, Oechel WC, Morison J, Zhang T, Barry RG (2000) Observational evidence of recent change in the northern high latitude environment. Clim Change 46:159–207

SFS (1971) Svensk författningssamling Notisum livsmedelslag. http://www.notisum.se/rnp/sls/ lag/19710511.htm. Accessed 12 Jul 2017

Shannon G, McKenna MF, Angeloni LM, Crooks KR, Fristrup KM, Brown E, Warner KA, Nelson MD, White C, Briggs J, McFarland S (2016) A synthesis of two decades of research documenting the effects of noise on wildlife. Biol Rev 91(4):982–1005

Sheehy T et al (2015) Traditional food patterns are associated with better diet quality and improved dietary adequacy in Aboriginal peoples in the Northwest Territories, Canada. J Hum Nutr Diet 28(3):262–271

Siebert C, Richardson J (2011) Food ark: preserving heirloom seeds and breeds is crucial if we are to feed our hungry world. Natl Geogr 220(1):108–131

Sing CF, Orr JD (1976) Analysis of genetic and environmental sources of variation In serum cholesterol in Tecumseh, Michigan. Ill-Identification of genetic effects using 12 poly morphic genetic blood marker systems. Am J Hum Genet 28:453–464

Sjölander P (2011) What is known about the health and living conditions of the indigenous people of northern Scandinavia, the Sami? Glob Health Action 4. https://doi.org/10.3402/gha. v4i0.8457

SLICA (2015) Survey of living conditions in the Arctic. https://intranett.uit.no/Content/432492/ Popul%C3%A6rvitenskapelig%20tidsskrift%20_%20Senter%20for%20Saamisk%20helse-forskning%202015.pdf. Accessed 10 Jan 2017

Snyder J (2007) Tourism in the polar regions: the sustainability challenge. UNEP/Earthprint, 2007

Soininen L, Järvinen S, Pukkala E (2002) Cancer incidence among Saami in Northern Finland, 1979–1998. Int J Cancer 100:342–346

Solstad JK (2012) Samisk Språkundersøkelse. NF-rapport nr.7/2012. In: Solstad JK (ed.). Nordlands Forskning. Retrieved 29 March 2014 from: http://nordlandsforskning.no/files/ Rapporter_2012/Rapport_07_12.pdf. Accessed 1 June 2017

Sorvali J (2015) Climate strategy work in the Barents region. http://www.climatesmart.fi/wp-content/uploads/2015/09/Climate-strategy-work-in-the-Barents-area-9.2015-english.pdf. August 2015

Staalesen A (2015) Consumer price hike in the Barents Russia." The Independent Barents Observer. December 2015. http://www.patchworkbarents.org/node/143. Accessed 23 Sep 2016

Staalesen A (2016) 26 years on, Russian death clouds still descend on Norwegian borderlands. The Barents observer. August, 4, 2016. http://thebarentsobserver.com/ecology/2016/08/26-years-russian-death-clouds-still-descend-norwegian-borderlands. Accessed 9 Nov 2016

Stockholm Convention (2004) Stockholm convention on persistent organic pollutants. https://www.un.org/press/en/2004/unep204.doc.htm. Accessed 26 July 2017

Storting (2013) The EEA Agreement and Norway's other agreements with the EU. https://www.regjeringen.no/globalassets/upload/ud/vedlegg/europa/nou/meldst5_ud_eng.pdf. Accessed 15 July 2017

Survival International (2016) The Nenets of Siberia. http://www.survivalinternational.org/photo-stories/3198-the-nenets-of-siberia. Accessed 7 Nov 2016. "Sweden." Grantham Research Institute on Climate Change and the Environment

Sverige Regering (2015) Ministry of Enterprise and Innovation. http://www.government.se/government-of-sweden/ministry-of-enterprise-and-innovation/. Accessed 15 July 2017

Szczepański M (2015) Economic impact on the EU of sanctions over Ukraine conflict. European Parliament, 2015

Taagepera R (2011) The Finno-Ugric republics and the Russian state. Routledge, New York

Tait H (2006) Aboriginal peoples survey, 2006: Inuit health and social conditions. Ottawa, Ontario: Statistics Canada (December 2008). Available: http://tinyurl.com/265p7ur. Accessed 16 April 2017

TASS (2016) Russian food embargo hits Finland, Norway & Lithuania worst. Russia & India report. September 9, 2016. https://in.rbth.com/news/2016/09/09/russian-food-embargo-hits-finland-norway-lithuania-worst_628423. Accessed 8 Nov 2016

Tchounwou PB, Yedjou CG, Patlolla AK, Sutton DJ (2012) Heavy metal toxicity and the environment. In: Molecular, clinical and environmental toxicology. Springer, Basel, pp 133–164

TEM (2017) Työ- ja elinkeinoministeriö. Finnish ministry of economic affairs and employment. https://tem.fi/en/search/-/q/food%20control%20of%20animal%20origin. Accessed 18 Aug 2017

The Climate Doctrine of the Russian Federation (2009) Climate observer. https://climateobserver.org/wp-content/uploads/2014/09/Climate-Doctrine.pdf. Accessed 18 Jul 2018

Tirado MC, Clarke R, Jaykus LA, McQuatters-Gollop A, Frank JM (2010) Climate change and food safety: a review. Food Res Int 43(7):1745–1765

Tong S, Mather P, Fitzgerald G, McRae D, Verrall K, Walker D (2010) Assessing the vulnerability of eco-environmental health to climate change. Int J Environ Res Public Health 7:546–564

Turunen MT, Rasmus S, Bavay M, Ruosteenoja K, Heiskanen J (2016) Coping with difficult weather and snow conditions: reindeer herders' views on climate change impacts and coping strategies. Clim Risk Manag 11:15–36

UIT (1999) Universitetet I Tromso (UIT), University Library of Tromso. http://www.ub.uit.no/northernlights/eng/pomor.htm. Accessed 8 May 2017

UN (2009) United Nations Human Security Unit. http://www.un.org/humansecurity/sites/www.un.org.humansecurity/files/human_security_in_theory_and_practice_english.pdf. Accessed 18 May 2017

UN (2017) United Nations Sustainable Development Goals. http://www.un.org/sustainabledevelopment/blog/2017/06/world-population-projected-to-reach-9-8-billion-in-2050-and-11-2-billion-in-2100-says-un/. Accessed 16 June 2017

UN IASG (2014) United Nations Inter agency support group report on indigenous peoples' issues. Land, territories and resources, June 2014 http://www.un.org/en/ga/president/68/pdf/wcip/IASG%20Paper%20_%20Lands%20territories%20and%20resources%20-%20rev1.pdf. Accessed 15 July 2017

UNCLOS (1982) The United Nations convention on the Law of the Sea. http://www.un.org/depts/los/convention_agreements/texts/unclos/unclos_e.pdf

UNDP (1994) UNDP Human Development Report of 1994 and the Human Security Unit. United Nations Development Programme, New York

UNECE (1998) United Nations Economic Commission for Europe. http://www.unece.org/info/ece-homepage.html. Accessed 17 Jul 2017

UNECE (2016) United Nations Economic Commission for Europe. https://www.unece.org/trans/main/dgdb/ac2/ac2nwdc_2016.html. Accessed 16 Jul 2017

UNEP (2014) Plastic waste causes financial damage of US$13 billion to marine ecosystems each year as concern grows over microplastics. United Nations Environmental programme press release. 23rd June 2014. http://www.unep.org/newscentre/plastic-waste-causes-financial-damage-us13-billion-marine-ecosystems-each-year-concern-grows-over. Accessed 21 June 2017

UNESCO (1995) World commission of culture and development, our creative diversity, report submitted to UNESCO and the United Nations in November 1995 Paris. http://unesdoc.unesco.org/images/0010/001016/101651e.pdf. Accessed 14 July 2017

UNFCC (1992) "United Nations framework convention on climate change." 1771 UNTS 107/1994, ATS 2/31 ILM 849, United Nations Secretary General, 1992

UNFSA (1995) U.N. Conference on Straddling Fish Stocks and Highly Migratory Fish Stocks, July 24–Aug. 4, 1995, Agreement for the Implementation of the United Nations Convention on the Law of the Sea of 10 December 1982 Relating to the Conservation and Management of Straddling Fish Stocks and Highly Migratory Fish Stocks, U.N. Doc. A/Conf. 164/37 [hereinafter UNFSA]. Available at http://www.un.org/depts/los/convention_agreements/texts/fish_stocks_agreement/CONF164_37.htm. Accessed 07 July 2017

UNGA (2007) United Nations general assembly, "United Nations declaration on the rights of indigenous peoples." A/RES/61/295, October, 2, 2007

UNPFII (2009) The State of the World's Indigenous Peoples. New York, United Nations Department of Economic and Social Affairs, Secretariat of the Permanent Forum on Indigenous Issues, pp. 250. www.un.org/esa/socdev/unpfii/documents/sowip_web.pdf. Accessed 17 July 2017

UNRIC (2016) United Nations regional information centre for Western Europe, "Individual vs. Collective Rights." http://www.unric.org/en/indigenous-people/27309-individual-vs-collective-rights. Accessed 25 Apr 2016

USAID (2013) Agency for international development (USAID). Integrating rule of law and global development: food security, climate change, and public health. Agriculture of the Republic of Komi. Republic of Komi: Official Portal. http://rkomi.ru/en/left/info/agro/. Accessed 7 Nov 2016

Usher PJ (1995) Communicating about contaminants in country food: The experience in aboriginal communities. Research Department, Inuit Tapirisat of Canada, Ottawa

Vahid F, Zand H, Nosrat-Mirshekarlou E, Najafi R, Hekmatdoost A (2015) The role of dietary bioactive compounds on the regulation of histone acetylases and deacetylases: a review. Gene 562(1):8–15

Valente FLS (2014) Towards the full realization of the human right to adequate food and nutrition. Development 57(2):155–170

Välkky E, Nousiainen H, Karjalainen T (2008) Facts and figures of the Barents forest sector. Working papers of the Finnish Forest Research Institute, 78. Available from: http://www.metla.fi/julkaisut/workingpapers/2008/mwp078.htm. Accessed 04 May 2017

van Oort B, Bjørkan M, Klyuchnikova EM (2015a) Future Narratives for two locations in the Barents region. Workshop report for scenario-building workshops in Kirovsk, June 9th, 2015 and Bodø, August 25th, 2015. CICERO Report 2015:06. CICERO, Oslo, Norway

Van Oort B et al (2015b) Future narratives for two locations in the Barents region. CICERO Report, December, 2015

Van Oostdam J, Donaldson SG, Feeley M, Arnold D, Ayotte P, Bondy G et al (2005) Human health implications of environmental contaminants in Arctic Canada: a review. Sci Total Environ 351:165–246

Van Oostdam J, Gilman A, Dewailly E, Usher P, Wheatley B, Kuhnlein H et al (1999) Human health implications of environmental contaminants in Arctic Canada: a review. Sci Total Environ 230(1-3):1–82

Van Sebille E, Matthew HE, Froyland G (2012) Origin, dynamics and evolution of ocean garbage patches from observed surface drifters. Environ Res Lett 7(4):044040

Vanderberg R et al (2015) Russian federation: food and agricultural import regulations and standards – narrative. FAIRS Country Report, December 15, 2015

VanderZwaag D (1998) International Commons 9 YB International Encyclopaedia of Laws Law at 272. In: Nowlan L (ed) Arctic legal regime for environmental protection, No. 44. International Union for Conservation of Nature IUCN, Gland, p 24

Vapnek J, Spreij M (2005) Perspectives and guidelines on food legislation, with a new model food law. Food and Agriculture Organization of the United Nations, Rome

Vassilieva E (2015) "Hot as Barents!" Barents Observer. January, 27, 2015. http://barentsobserver.com/en/nature/2015/01/hot-barents-27-01. Accessed 11 Nov 2016

Vermeulen SJ, Campbell BM, Ingram JS (2012) Climate change and food systems. Ann Rev Environ Resour 37:195–222

Verschuren WM, Jacobs DR, Bloemberg BP, Kromhout D, Menotti A, Aravanis C, Blackburn H, Buzina R, Dontas AS, Fidanza F, Karvonen MJ (1995) Serum total cholesterol and long-term coronary heart disease mortality in different cultures: twenty-five—year follow-up of the seven countries study. JAMA 274(2):131–136

Vidal J (2016) Mining threatens to eat up northern Europe's last wilderness. September 3, 2016. https://www.theguardian.com/environment/2014/sep/03/mining-threat-northern-europe-wilderness-finland-sweden-norway. Accessed 28 Nov 2016

Viires A (1993) The Red Book of the Peoples of the Russian Empire. ISBN 9985-936922. http://www.eki.ee/books/redbook/veps.shtml. Accessed 7 Nov 2016

Visitbodo (2017) Official travel site for Bodo. www.visitbodo.com. Accessed 15 June 2017

Visitrovaniemi (2017) The pure taste of Rovaniemi. http://www.visitrovaniemi.fi/love/local-food/. Accessed 10 May 2017

Volder EC, Li YF (1995) Global usage of selected persistent organochlorines. Sci Total Environ 160/161:201–210

VTT (2012) Low Carbon Finland 2050. VTT clean energy technology strategies for society. http://www.vtt.fi/sites/lowcfin/en. Accessed 14 July 2017

Wartiainen (2007) Climate in change: nature and society challenges for the Barents region. http://www.bioforsk.no/ikbViewer/Content/96985/BW07_engelsk_nett%20(2).pdf. Accessed 15 Jul 2017

Wartiainen I (2007) Global climate change results in large regional variations. Barents Watch: Climate in Change Nature and Society Challenges for the Barents Region, 2007

Wärtsilä (2017) Wärtsilä joins seabin project in the battle against ocean plastics. https://www.wartsila.com/media/news/10-02-2017-wartsila-joins-seabin-project-in-the-battle-against-ocean-plastics. Accessed 29 May 2017

Wassman P, Reigstad M, Haug T, Pavlov D (2006) Food webs and carbon flux in the Barents Sea. Prog Oceanogr 71(2–4):232–287. Available from: https://www.researchgate.net/publication/223542984_Food_webs_and_carbon_flux_in_the_Barents_Sea. Accessed 9 May 2017

WCRF (2007) World cancer research fund: food, nutrition, physical activity and the prevention of cancer a global perspective. Publisher is the American Institute for Cancer Research (AICR), Washington DC. American Institute for Cancer Research

Wegren SK, Nikulin AM, Trotsuk I (2016a) The Russian variant of food security. Problems of post-communism. https://doi.org/10.1080/10758216.2016.1163229. [Taylor & Francis Online]. Accessed 25 Apr 2017

Wegren SK, Nilssen F, Elvestad C (2016b) The impact of Russian food security policy on the performance of the food system. Eurasian Geogr Econ 57(6):671–699

Welsch RL, Vivanco LA (2015) Foodways: finding, making, and eating food. Cultural anthropology: asking questions about humanity. Oxford UP, New York, pp 157–183

Wennberg M, Bergdahl IA, Stegmayr B, Hallmans G, Lundh T, Skerfving S et al (2007) Fish intake, mercury, long-chain n-3 polyunsaturated fatty acids and risk of stroke in northern Sweden. Br J Nutr 98(5):1038–1045

WFC (1974) World food conference, Rome, 5–16 November 1974. Communication from the Commission to the Council. SEC (74) 4955 final, 9 December 1974

WFFS (2001) World Forum on Food Sovereignty Final Declaration, 3rd - 7th September, 2001. Havana, Cuba

Wheeler T, von Braun J (2013) Climate change impacts on global food security. Science 341(6145):508–513. https://doi.org/10.1126/science.1239402

White P, Michelck J, Lerner J (2007) Linking conservation and transportation: using the State Wildlife Action Plans to Protect Wildlife from Road Impacts. Defenders of Wildlife Report.

http://www.defenders.org/sites/default/files/publications/linking_conservation_and_transportation.pdf. Accessed 12 July 2017

WHO (2016) World Health Organization, "Climate change and human health: Biodiversity." http://www.who.int/globalchange/ecosystems/biodiversity/en/. Accessed 25 Nov 2016

Wideback A (2011) Food and Agricultural import regulations and standards – narrative, December 2011

Wiessner S (2009) The United Nations declaration on the rights of indigenous peoples. In: The diversity of international law. Brill, Netherlands, pp 343–362. http://www.converge.org.nz/pma/DRIPGA.pdf. Accessed 15 July 2017

Willows ND (2005) Determinants of healthy eating in aboriginal peoples in Canada: the current state of knowledge and research gaps. Can J Public Health 96(3):S32–S36

Windfuhr M, Jonsén J (2005) Food Sovereignty: Towards democracy in localized food systems. http://agris.fao.org/agris-search/search.do?recordID=GB2013202621. Accessed 30 July 2017

WIPO (2000) World Intellectual Property Organization. http://www.wipo.int/pressroom/en/briefs/tk_ip.html. Accessed 23 May 2017

WIPO (2009) World Intellectual Property Organization. http://www.wipo.int/tk/en/igc/. Accessed 23 May 2017

Woodley E et al (2006) Cultural indicators of Indigenous Peoples' food and agro-ecological systems. SARD Initiative commissioned by FAO and the International India Treaty Council, 1–104

World Bank (2015) World Bank indicator. http://data.worldbank.org/indicator/AG.LND.FRST.ZS. Accessed 12 July 2017

WTO-SPS (2015) World Trade Organization - Sanitary and Phytosanitary - Commission on Phytosanitary Measures Report CPM 2015/INF/07, Rome, Italy. https://www.ippc.int/static/media/files/publications/en/2015/02/12/cpm_2015_inf_07_wto_sps_report_2015-02-12.pdf. Accessed 14 July 2017

WWF (2016) World Wide Fund for Nature, "Arctic Oil and Gas." http://wwf.panda.org/what_we_do/where_we_work/Arctic/what_we_do/oil_gas/. Accessed 1 Apr 2016

Yakovleva NV (2005) Tradicionnoe pitanjie zhitelei severa (Traditional nutrition of residents of North). Lomonosov Pomor State University, Arkhangelsk, p 244. [in Russian]

Young OR (2016) The shifting landscape of Arctic politics: implications for international cooperation. Polar J 6(2):209–223

Zewdie A (2014) Impacts of climate change on food security: a literature review in Sub Saharan Africa. J Earth Sci Clim Change 5: 225. doi:https://doi.org/10.4172/2157-7617.1000225. At: https://www.omicsonline.org/open-access/impacts-of-climate-change-on-food-security-a-literature-review-in-sub-saharan-africa-2157-7617.1000225.php?aid=30963. Accessed 7 July 2017

Printed in the United States
By Bookmasters